TECHNICAL GUIDE
Spitfire

TECHNICAL GUIDE
Spitfire

Edward Ward

This Amber edition published in 2023

Copyright © 2023 Amber Books Ltd

All rights reserved. No part of this publication may be reproduced, stored in a retrieval system, or transmitted in any form or by any means, electronic, mechanical, photocopying, recording, or otherwise, without prior written permission of the copyright holder.

Published by
Amber Books Ltd
United House
North Road
London N7 9DP
United Kingdom
www.amberbooks.co.uk
Instagram: amberbooksltd
Facebook: amberbooks
Pinterest: amberbooksltd

Editor: Michael Spilling
Designer: Mark Batley
Picture researcher: Terry Forshaw

ISBN: 978-1-83886-324-1

Printed in China

Contents

Introduction	6
Developing a Prototype	10
Mks I–III: Battle of Britain	26
Mks V–VI: Spitfires of the World	48
Mks VII–XVI: Ultimate Merlin Marks	62
Mks XII–XVIII: Griffon Engine Variants	80
Photo Reconnaissance Models	90
Mk 21 to the Spiteful	104
The Seafire	114
Index	126
Picture Credits	128

Introduction

Just before 5 o'clock in the evening on 5th March 1936, a small group of engineers watched the prototype of a new fighter aircraft take off from Eastleigh aerodrome in the south of England for its maiden flight.

After a satisfactory eight-minute flight, the aircraft landed and taxied back to the hangar. The pilot, Joseph 'Mutt' Summers, climbed from the cockpit saying, 'I don't want anything touched.' Later commentators took this to mean that the aircraft was immediately perfect. However, what Summers actually meant was that there were no corrections or modifications he wanted made before he flew it again. The aircraft was the Supermarine Type 300, and whilst those present were filled with a cautious optimism regarding the fortunes of their new fighter, it is unlikely that anyone present would have dared imagine that this was destined to become the most famous British aircraft of all time. So important would it prove to the British perception of the war effort that this aircraft was to become iconic even as it tangled in the skies with the Axis forces.

Spitfire spirit

The aircraft's name, 'Spitfire', would be invoked to describe the defiant mood of the nation as the 'Spitfire spirit', and borrowed to encourage donations to the war effort by various 'Spitfire funds'. Later it would become utterly synonymous with Britain's 'finest hour' during the Battle of Britain, the 'Spitfire summer'. Today the word 'Spitfire' is more readily understood as a World War II fighter aircraft than as its original meaning: a woman with a fiery temper. Reginald Mitchell, the aircraft's designer, allegedly said, 'It's just the sort of bloody silly name they would think of', his preference being for a name guaranteed to strike fear into the heart of any enemy: 'Shrew'.

But this was just one of several seeming dichotomies of the Spitfire programme, for here was an aircraft that was complex, time-consuming and expensive to build yet was mass-produced by largely unskilled labour in greater numbers than any other British aircraft. This was an aircraft thought to be conceptually obsolete even before its first flight but which turned out to be the only Allied fighter to remain in production and service from the first day of the war to the last; this was an aeroplane that shot down more enemy aircraft than any other Allied fighter, yet whose first victims were RAF Hurricanes; and this was a machine explicitly designed as a short-range interceptor that became the Allies' premier long-range strategic reconnaissance asset. Even the event with which it has become irrevocably linked, the Battle of Britain, saw it play a supporting role, admittedly a critically important one, to the considerably more numerous Hurricane.

Legendary adaptability

Even without the subsequent legendary status it attained, the story of the Spitfire is a compelling one: a thoroughbred developed from a series of record-setting racing aircraft by a young dying genius who would not live to see his creation's incredible triumph. The Spitfire was an aircraft that married arguably the world's finest aero engine, the Merlin, with the most aerodynamically advanced high-speed airframe yet devised for a military aircraft. The Merlin was not designed exclusively for the Spitfire and would provide yeoman service in a vast array of aircraft types, but its synthesis with the Spitfire would prove a particularly happy alliance, rivalled

Opposite: FILM STAR QUALITY Both AR213 (furthest from camera), a Mk Vb built in 1941 by Westland and MH434, a Mk IXB built at Castle Bromwich in 1943, starred in the 1968 film *Battle of Britain*. MH434 subsequently appeared in *Dark Blue World* in 2001 and AR213 in *Dunkirk* in 2017.

INTRODUCTION

only perhaps by the somewhat later North American Mustang.

As the war progressed, an unsuspected adaptability would see the Spitfire excel in roles for which it had never been envisioned or intended. The defensive bomber interceptor of 1940 was later pressed into service as an escort fighter, a ground attack aircraft and a stupendously successful photo-reconnaissance aircraft. It was even used as the most unlikely transport aircraft of the war in a bizarre beer delivery role following the Allied invasion of Europe. Most radically of all, the Spitfire went to sea as a carrier aircraft, a process that was not an unequivocal success – especially when compared to its other manifold roles – but eventually even the Seafire matured into an effective and useful naval aircraft.

Widespread service

Its matchless qualities saw the Spitfire flown by a cosmopolitan variety of Allied air arms during the war, including the Soviet Union and USA, and it became the most numerous and successful foreign combat aircraft in the American inventory. Other nations such as Australia, Canada, France and Italy all flew the Spitfire operationally

MODIFICATION XXX
Likely the most unconventional role for the wartime Spitfire, delivering beer to the troops in France. The external fuel tank filled with beer was fitted under the wing, though later beer flights dispensed with the tanks and simply carried a barrel under each wing.

during the conflict. Even the German Luftwaffe saw fit to utilize a Spitfire on test work. Long after the war that it was designed to fight, the Spitfire carried on serving in the inventories of a swathe of nations, scoring its last few 'kills' in a confused and confusing three-way engagement between the

INTRODUCTION

RAF, the Israeli Hel Ha-avir and the Egyptian Air Force – all three equipped with Spitfires.

The appeal of the Spitfire as both engineering masterpiece and cultural artefact endures: scale replicas are available of every variant as model kits and die-cast models. Even flyable scale homebuilt aircraft are available for the aspiring Spitfire pilot (with DIY skills). Ever more Spitfires join the ranks of preserved aircraft on the airshow circuit every year: at the time of writing, 54 Spitfires and Seafires are in airworthy condition worldwide, more than when the RAF retired its Spitfires in 1955 just short of twenty years after the prototype first flew.

The following chapters trace the story of this remarkable machine from the pretty but largely unremarkable fighter prototype of 1936 via the myth-making events of 1940 through to the blisteringly fast Griffon-engined later models.

Details of the Spitfire's development and service follow, which will explain its huge success as a military aircraft, though this only partly accounts for the Spitfire's enduring fame in the popular imagination – a fame that one struggles to believe would still be the case had it been named the Shrew. Few aircraft can be considered truly iconic, but the Spitfire is undoubtedly one of them.

POSTWAR ACTION
The active career of the Spitfire continued long after the end of World War II. This Egyptian Mk IX was shot down by Israeli forces during the 1948 Palestine War.

Developing a Prototype

Remarkably, the Spitfire was both RJ Mitchell and the Supermarine aircraft company's first fighter design to enter production, Mitchell's Supermarine designs up until this point having been almost exclusively seaplanes and flying boats. Nonetheless, such an advanced machine had not sprung up out of nowhere; by the time of the Spitfire's first flight in 1936, Mitchell had gained a wealth of experience designing high-speed aircraft intended not for combat but for racing.

FIRST SHOWING

K5054's first public appearance came at a display of Vickers types before 300 invited guests at Eastleigh on 18 June 1936. Other aircraft on show were K4049, the B.9/32 bomber prototype (later ordered by the RAF as the Wellington); K7556, a pre-production Wellesley bomber; and K5780, the ninth production-standard Walrus amphibian.

DEVELOPING A PROTOTYPE

NOEL PEMBERTON-BILLING
A brilliant inventor, Pemberton-Billing's reputation was compromised by his increasingly extremist views and his willingness to resort to fabricating evidence to support them.

In 1913, on the banks of the River Itchen in Southampton, Pemberton-Billing Ltd – the company that would later become Supermarine – was founded by the highly prolific inventor and publisher Noel Pemberton-Billing. Initially, the company traded motorboats and was highly profitable, allowing Billing to build a few of his unsuccessful flying boat designs. In the lead-up to the outbreak of war in August 1914, Pemberton-Billing Ltd just managed to stave off bankruptcy, in part by leasing its premises to Sopwith for the construction and testing of that company's wonderfully named 'Bat Boat'. Subsequently Pemberton-Billing found work subcontracting for other manufacturers, and the needs of war meant that funds were now readily available for experimental designs and prototypes.

As the war progressed, Billing was less and less involved in the day-to-day running of the firm,

SUPERMARINE NIGHTHAWK
Impressive but useless, the Nighthawk was the first product of the newly created Supermarine Aviation Works Ltd. Huge, unwieldy and slow, the Nighthawk was the very antithesis of its distant descendant, the Spitfire.

and in 1916, he was elected as an MP. After his election Pemberton-Billing was extremely vocal in Parliament in his criticism of the government-owned Royal Aircraft Factory and the aircraft that it produced, campaigning for it to be shut down. Realizing that this might prejudice their relationship with the Government and potentially lead to the loss of future orders, the management of Pemberton-Billing Ltd sought to distance themselves from Billing himself, with the other directors of the company buying back Billing's shares, the majority by designer and factory manager Hubert Scott-Paine, and renaming the firm the Supermarine Aviation Works Ltd in June 1916. The name 'Supermarine' was taken from the telegraph address of the company, Billing coined it as he figured that if a submarine was effectively a boat that could travel under the ocean, then a 'Supermarine' was a boat that could fly over it.

Supermarine Nighthawk
The company's separation from Pemberton-Billing would prove sensible as he became increasingly extremist and fantasist in his political views, and his aircraft designs proved ill-conceived. Nonetheless, the first aircraft destined to be built by Supermarine was one of Billing's pet projects, the 'Nighthawk', a remarkable but impractical machine intended to find and destroy Zeppelins. The Nighthawk was one of the very few quadruplanes to have been built and flown and was, in some respects, a harbinger of things to come. It featured a powerful searchlight in the nose, a very heavy armament and a fully enclosed heated cabin for the crew, complete with a rest bunk. This was deemed necessary as the Nighthawk had a remarkable design endurance of some 14 hours.

Unfortunately, all this equipment was very heavy, and the large, four-winged Nighthawk possessed a huge amount of built-in drag. With only two 100hp Anzani engines to haul it along, testing revealed that it was unable to climb to the height at which

Zeppelins operated and, with its woeful 100km/h (64mph) top speed, it was comfortably outrun by them as well. Nonetheless, the Nighthawk could stake its place in aviation history as the world's first purpose-designed nightfighter. More significantly perhaps, it was also the first project on which a young draughtsman named Reginald Joseph Mitchell worked, better known to his colleagues as 'RJ'.

New designer

RJ Mitchell, like many of his contemporaries, had learned engineering through the railways, in Mitchell's case as an apprentice at Kerr Stuart locomotive works. However, from an early age, his true interest lay in aviation. Mitchell joined Supermarine in 1916 after answering a newspaper advert, arriving in time to design the tailplane of the Nighthawk. Mitchell's obvious talent saw him promoted within a year to be the assistant of Hubert Scott-Paine, Supermarine's then owner, and by 1918 he had been further promoted to the role of assistant to the works Manager. Following the departure of chief designer Walter Hargreaves from Supermarine in 1919, the 24-year-old Mitchell was chosen to

Opposite: RJ MITCHELL
A brilliant designer, Vickers' takeover of Supermarine in 1928 was widely believed to be a means to secure Mitchell's services. Ultimately, his greatest work would be undertaken while he died of cancer.

replace him – a remarkably young age to be appointed to such a position. Mitchell would remain in the post for the rest of his working life.

1919 also saw the return of the Schneider Trophy seaplane contest, which had been discontinued for the duration of World War I and would prove to be instrumental in raising the profile of the Supermarine company and cementing Mitchell's reputation as a designer of high-speed aircraft. The 'Coupe d'Aviation Maritime Jacques Schneider' had been created by French industrialist and pioneering aviator Jacques Schneider to promote the development of practical seaplanes, which in 1913, when the first competition was held, seemed to offer greater promise than their land-based counterparts. In a world which had not yet seen the development and construction of manifold airports, Schneider reasoned, the development of larger and faster aircraft was best suited to maritime aircraft, which could land and take off from any sufficiently large body of water. Ultimately the proliferation of aerodromes and landing grounds as a result of World War II rendered Schneider's vision obsolete. However, during the interwar period, the Schneider Cup developed a prestige unmatched by any other air-racing competition and provoked enormous efforts from aircraft companies and nation states to win it. The rules of the competition required contestants to prove adequately seaworthy, but this aspect of the Schneider Cup soon became little more than a formality before the all-important flat-out speed race.

Race time

The first Supermarine aircraft to race for the Schneider Cup was the Sea Lion, a development of the Supermarine Baby, a prototype single-seat fighter flying boat that had first flown in February 1918 but had not entered production. When the Sea Lion first appeared, it was the world's smallest and fastest flying boat. As such, it was an excellent basis for a potential racer. Although designed by Walter Hargreaves, Mitchell would undoubtedly have worked on the Sea Lion though the extent of his contribution is not now known. Powered by an example of the new 450hp Lion engine that had been loaned to Supermarine by Napier, and which would become one of the most successful aero engines of the 1920s and 1930s, the Sea Lion was fast for its era at 237km/h (147mph), and hopes were high for its performance during the race.

The 1919 Schneider meeting was held at Bournemouth and proved to be a chaotic fiasco culminating in an apparent Italian victory for the Savoia S.13. Unfortunately for the Italians, the pilot Guido Jannello had missed the course marker in thick fog and had actually been flying a much shorter circuit. Although the error was both innocent and understandable due to the fog, the

DEVELOPING A PROTOTYPE

Italian aircraft was disqualified and the whole contest declared void because no other aircraft had managed to complete the course. Basil Hobbs, the pilot of the Sea Lion, had similarly been unable to locate the course marker boat in the fog but had decided to alight on the water to see if he could find it. Unsuccessful, he decided to fly back to the start point anyway. Unfortunately, the Sea Lion struck an unknown object in the water on take-off, tearing a large hole in the hull. When Hobbs attempted to land at Bournemouth, the hole caused the Sea Lion to capsize – thankfully without injury to Hobbs, who was thrown clear.

Seal Lion II

Despite this ignominious end to the Sea Lion's racing career, the aircraft had made an impression with its speed, providing much needed attention for the Supermarine company. For the next two years, no British entry was made to the Schneider Cup, and Italian aircraft won both times. The Schneider rules were such that if any nation won it three times in a row, they would be considered the permanent victors, and no further competitions would be held. Italy were just one win away from this eventuality, and Hubert Scott-Paine, managing director of Supermarine, believed they had the aircraft to win the race for Britain. The aircraft was the Sea Lion II, of the same basic layout as the original Sea Lion, but faster, stronger and very

manoeuvrable. This aircraft was a racing derivative of the Sea King, a proposed amphibious fighter intended to operate from an aircraft carrier or the water. This time the original aircraft was entirely Mitchell's design, and he oversaw its modification to a racing machine. The principal change was the substitution of the Sea King's Beardmore engine for the latest development of the Napier Lion, delivering some 50 per cent more power than the original power

SCHNEIDER TROPHY

Likely the most prestigious air race in history, the lure of the Schneider trophy prompted feverish action on the part of both aircraft designers and national governments. The current piston-engined seaplane speed record is still held by an aircraft designed for the Schneider Trophy.

unit. The amphibious landing gear was removed, and alterations were made to the basic structure to allow the aircraft to handle the power of

DEVELOPING A PROTOTYPE

its new engine. Christened the Sea Lion II, the superficial resemblance between the new Supermarine racer and the Sea Lion of 1919 effectively deceived observers at the competition of its true capability, and when pilot Henri Biard opened up the Sea Lion II to full throttle, its speed performance came as a rude shock to the Italian team. Biard won the speed contest by over two minutes. Further tweaks to the basic design yielded the Supermarine III for the following

SUPERMARINE SEA LION II
Victor of the 1922 Schneider Trophy race, the Sea Lion II was the first British aircraft to win an international competition after the end of the Great War. This success generated much-needed interest in Supermarine.

year's competition, but this aircraft was comfortably outpaced by the Curtiss CR-3 floatplanes of the US Navy and finished a distant third at an average speed some 32km/h (20mph) behind the American machines. Supermarine would have to produce something spectacular in order to defeat them.

Somewhat surprisingly, the CR-3, which Curtiss had developed purely for racing, was doubly influential on the Spitfire. As well as provoking Mitchell to investigate the cantilever monoplane as a means of outpacing it, its engine, the V-12 Curtiss D-12 featuring a cast aluminium cylinder block, was to prove enormously influential, and directly inspired Rolls-Royce to produce their own cast-block V-12 engine: the Kestrel. All

subsequent Rolls-Royce V-12 engines, including the Merlin and Griffon, owe their existence to the revolutionary D-12. This was all to come later though, and Supermarine once again turned to the powerful and reliable Napier Lion for their next racer, the S.4.

Supermarine S.4 and S.5 racers
The S.4 was conceived and built in 1925 but could pass for an aircraft designed a decade later. A remarkably elegant aircraft, it was the first to truly show Mitchell's flair for producing an exceptionally clean, streamlined airframe. Although it appeared within a couple of years of the Sea Lion III, the white-painted S.4 looked like it had emerged from another epoch. Its most revolutionary and

17

SUPERMARINE S.5
The victorious 1927 Schneider team, with a jubilant Mitchell in the centre wearing the dark jacket and white trousers. Winning pilot Flt Lt (later Air Vice Marshal) Sidney 'Pebbler' Webster is standing on the float.

DEVELOPING A PROTOTYPE

controversial feature was its wing. Eschewing the drag-producing bracing wires and struts that were generally de rigueur for aircraft of this era, the S.4's wing was a self-supporting cantilever structure with a remarkably clean form. Supermarine's chief test pilot Henri Biard, who had triumphed in Naples with the Sea Lion II, was suspicious of the unbraced wing and, according to his autobiography, never felt entirely safe in the S.4. Nonetheless, within a month of its first flight, Biard had piloted the S.4 to a new seaplane speed record of 365km/h (227mph) that boded extremely well for the Schneider meeting.

As it turned out, the S.4 never got the chance to prove itself, as the aircraft crashed before the speed competition was held. Henri Biard escaped without serious injury despite the aircraft impacting the water at high speed. The result of the official investigation into the crash was that the aircraft had stalled, but there were suspicions about the unbraced wing, which Biard believed to be the cause. Today it is generally believed that the wing suffered from 'flutter' – oscillating due to aerodynamic forces twisting it out of shape – and Biard lost control because of this phenomenon. The truth will never definitively be known. It is notable, though, that Mitchell returned to a wire-braced wing for the successor to the S.4 – the S.5 – when it appeared in 1927. The S.5 dispensed with the bulky Lamblin radiators that contributed some 30 per cent of the total aerodynamic drag of the S.4 and replaced them

SUPERMARINE S.6B
Two S.6Bs were built, S1595, which won the 1931 Schneider Trophy, and S1596, seen here, which was flown by Flt Lt George Stainforth to a new world speed record of 655.67 km/h (407.5 mph) seventeen days later.

with surface-cooling radiators. Featuring a more powerful Napier Lion, S.5s finished first and second at that year's Schneider race.

It was clear, however, that the Lion was nearing the end of its development, and for higher speeds, a new engine of greater power would be required. Realizing this, engineers at Rolls-Royce schemed a new engine specifically for racing and derived from the Rolls-Royce Buzzard – itself a scaled-up Kestrel. The new engine was the Rolls-Royce 'R' (for

DEVELOPING A PROTOTYPE

'Racing') and first ran in 1929. Plans of the new engine were sent to Mitchell at Supermarine, and the new S.6 was designed to utilize it.

Supermarine S.6

The Supermarine S.6 thus united Mitchell's aerodynamic genius with Rolls-Royce's prowess in engine design, and the aircraft's relationship to the Spitfire is clear at a glance, especially in profile. Despite retaining the wire-braced wing, the new aircraft was even more finely streamlined than the S.5, and with the R engine producing around 1900hp, the S.6 romped home to win the 1929 contest at Calshot. For the first time, the winning speed was over 483km/h (300mph) – the 529km/h (329mph) average speed of the winning S.6 was faster than the absolute airspeed world record at the time. The following Schneider Trophy Rolls-Royce boosted the R to 2350hp, and with no other challenger ready for the race in time, the Supermarine S.6b completed the course unchallenged at 547km/h (340mph) and won the trophy outright for the UK. A few days later, the S.6b pushed the world airspeed record over 644km/h (400mph) for the first time. Examples of the 19 Rolls-Royce R engines that were constructed, including the same example used in the S.6B, were later used to power several successful land and water speed record vehicles. This included Malcolm Campbell's 'Bluebird' record-setting car and boat, and George Eyston's massive eight-wheeled 'Thunderbolt' car, which utilized two R engines to raise the land speed record to over 563km/h (350mph).

Type 224

Mitchell's triumph in the Schneider, however, was not matched by his first land-based fighter design. In late 1931, the Air Ministry issued specification F.7/30 for an all-metal fighter armed with four machine guns and a top speed of 314km/h (195mph). The Ministry's preferred engine was the Rolls-Royce Goshawk, which featured an unusual pressurized evaporative cooling system. Supermarine's response to F.7/30 was the Type 224, curiously enough nicknamed 'Spitfire' – Supermarine going so far as to request the name be reserved for it by the Air Ministry. The Type 224 featured an inverted gull wing similar to that later used by the Junkers Ju 87 'Stuka' and a fixed undercarriage enclosed by large trouser fairings. First flown in 1934, the Type 224's performance was disappointing. The aircraft was considerably slower than predicted, and its climb performance was particularly underwhelming. The highly unreliable nature of the evaporative cooling was also disappointing, with the pressurization system failing regularly. Even when it worked perfectly, the evaporative cooling system required large wing-mounted condensing cooling surfaces, which were likely to prove prohibitively vulnerable in combat. The F.7/30 specification was eventually fulfilled by the Gloster Gladiator, a conventional biplane with an air-cooled radial Bristol Mercury engine.

The Gladiator would become the last biplane fighter to enter service with the RAF and enjoy a distinguished career during the early years of World War II. Although superior to the Type 224, and the other competing designs

SUPERMARINE TYPE 224
The relatively uninspiring performance of the Type 224, the first 'Spitfire' and Supermarine's first single-seat fighter design, prompted Mitchell to rethink his ideas and directly led to the creation of the Supermarine Type 300.

21

DEVELOPING A PROTOTYPE

for F.7/30, it clearly represented the last of a dying breed, soon to become obsolete.

Mitchell, painfully aware that the Type 224 would not prove successful, began scheming changes that might yield a better fighter. It soon became clear that what was required, rather than a modification of the existing aircraft, was a new design. Taking the Type 224 as a starting point, Mitchell reduced the wingspan (by around 1.8m/6ft) and the thickness of the wing, ditching the inverted gullwing layout in the process. Flaps, a retractable undercarriage and an enclosed cockpit were added, and the tail surfaces were changed. With four machine guns fitted in the wings outboard of the propeller disc, the new design was referred to as the Type 300, and confident they had an excellent fighter in their hands, Supermarine approached the Air Ministry for funding. The Ministry was extremely keen and promptly supplied funding for the construction of a prototype, writing a new specification F.37/34 specifically to cover it. This was altered to F.10/35 when it became apparent that a four-gun armament would be insufficient and at least six (but preferably eight) machine guns would be required.

Merlin engine
Mitchell's biggest initial problem was power: the Goshawk was not being developed further due to the

DEVELOPING A PROTOTYPE

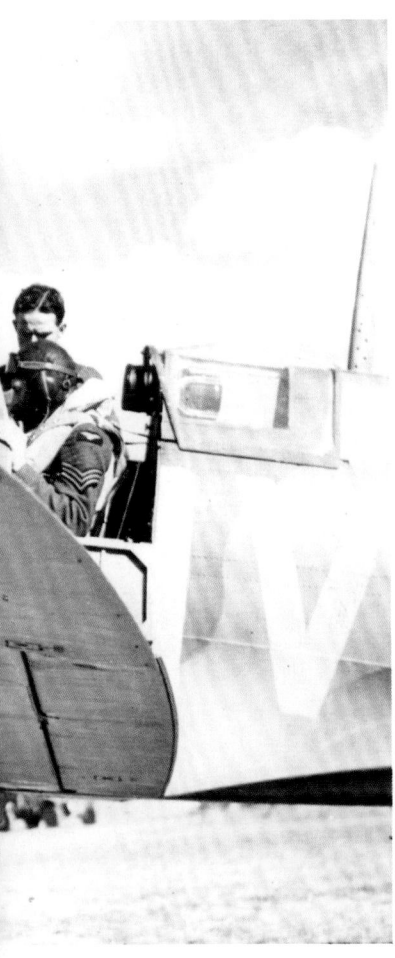

flaws in the evaporative cooling concept, and other engines in the same power class were few and far between. The 700hp Napier Dagger was considered for a time but whilst an improvement over the Goshawk, its power output increasingly appeared too modest. Luckily for Mitchell, Rolls-Royce had identified a need for a larger capacity engine than the Kestrel and, in the absence of any government funding, had been developing a new V-12 engine as

ELLIPTICAL WING
The most obvious external recognition point of the Spitfire is its elegant elliptical wing. Although fantastically efficient, it was time-consuming and expensive to mass produce. This 19 Squadron Mk IA is in the process of being rearmed at RAF Fowlmere in September 1940.

a private venture designated the PV12. The PV12 was a 27-litre displacement motor, first run in October 1933, that developed 800hp by 1934, with 1000hp expected soon. Thanks to the Schneider Trophy aircraft, Mitchell had an excellent relationship with Rolls-Royce, and the company reached an agreement for the PV12 – soon to be named 'Merlin' – to be fitted in Supermarine's new fighter. The combination of Merlin and Spitfire airframe would prove to be inspired, but the early development of the engine was far from plain sailing, with engines experiencing cracked cylinder heads, coolant leaks and excessive wear to the camshafts and main bearings. However, Rolls-Royce gradually eradicated the worst flaws of its new engine. In February 1935, an example of the Merlin took to the air for the first time in a modified Hawker Hart biplane.

Elliptical wing
Whilst the Merlin would power many other fighter aircraft, the other major development of the Spitfire, its elliptical wing, was unique and largely responsible for its lasting aesthetic appeal. As

well as improving the look of the aircraft, the wing was the key to the Spitfire's exceptional handling and probably the most aerodynamically advanced wing yet developed. This remarkable piece of design was primarily the work of a Canadian aerodynamicist called Beverly Shenstone who had come to Mitchell's team from the engineering giant Vickers after it acquired Supermarine in 1928. Shenstone was the first academically trained aerodynamicist to work at Supermarine and he brought a wealth of experience with him, as he had previously worked at Junkers in Germany and with Alexander Lippisch, pioneer of the delta wing, with whom Shenstone remained in contact after moving to the UK. Shenstone consulted with Lippisch about the Spitfire's wing as it progressed, which is somewhat ironic as Lippisch was heavily involved with German aeronautical development and would later design the 'Komet' rocket-propelled fighter for Messerschmitt.

Shenstone had also toured the American aviation industry in 1934, notably meeting Theodor von Karman, the Hungarian émigré aerodynamicist then working at Douglas, and was entranced by his work on laminar flow wings. One of the most important features of the Spitfire wing was its smoothness – Shenstone insisted on the use of flush rivets throughout the Spitfire's airframe, and although

23

DEVELOPING A PROTOTYPE

JOSEPH 'MUTT' SUMMERS
Summers joined the RAF in 1918 and became chief test pilot of Vickers in 1929. The Spitfire was just one of the 54 types that Summers took up on their first flights, an unbroken record, culminating in the turbojet Vickers Valiant nuclear bomber in 1951.

unable to attain true laminar flow, which was beyond the state of the art in the mid-1930s, this did result in a much less turbulent boundary layer than any of its contemporaries and therefore incurred less drag. As to the ellipse shape, or to be more accurate, the swept-forward ellipse, this was chosen primarily because it offered the best possible aerodynamics across as wide a speed and incidence range as possible. Turbulent, drag-inducing and lift-reducing air formulates primarily where high- and low-pressure airflows meet, principally along the trailing edge and the wingtip.

Because of this, modern airliners feature upswept winglets that effectively funnel spilled air off the wing's upper surface. By the 1920s, the pointed elliptical wing had been proven to result in less spillage and turbulence off the wingtip than could be achieved by a straight tip; it also possessed a smoother lift pattern over the entire wing.

The Spitfire's wing also possessed a slight twist along its length. Known as 'washout', both geometrically and helically, this feature reduced the tendency of air to flow spanwise, reducing lift, rather than straight over the wing front to rear. A cruder solution to this problem is the wing fence, which forces the air to conform to the wing's aerofoil profile. However, this also increases drag and decreases the lifting area. Introducing a slight wing twist maintains spanwise flow and adds very little drag, and Shenstone spent many hours working out just how much he could twist the wing before the onset of drag cancelled out the benefits of doing so.

Body and tail
The swept-forward nature of the ellipse meant that the main spar could be fixed at 90 degrees to the fuselage for maximum strength and the leading edge of the wing attached to the spar to form an immensely strong D-shaped box girder. Where the fuselage met the wing, the Spitfire possessed fuselage fillets, much larger than virtually all of its contemporaries, which Shenstone schemed to smooth

the join from wing to fuselage and reduce drag. By contrast, the Spitfire has a distinctively small tail. Tail surfaces impart significant drag to any airframe, and both Mitchell and Shenstone wanted to keep them as small as possible. The low wing loading of the elliptical wing meant that the effort required to control the aircraft was also comparatively low and the tail surfaces could be kept to a minimal size whilst retaining effectiveness - many pilots later remarked on the highly effective and sensitive elevators.

The application of rigorous scientific principles had produced a wing that retained the same level of efficiency at 80km/h (50mph) as at 805km/h (500mph) without resorting to high-lift devices yet possessed low wing loading, thus imparting excellent manoeuvrability and a docile stall with plenty of warning. The inevitable downside of this highly complex, aerodynamic marvel, made entirely of compound curves and with a different aerofoil section along its span, was that it was difficult and time-consuming to build: the Spitfire was a notably expensive fighter when compared to many of its contemporaries.

Type 300 prototype
Construction of the advanced Type 300 prototype, serial numbered K5054, was subject to infuriating delays. Although the aircraft was expected to be ready by November 1935, the first flight took place on 5 March 1936, four months after its

DEVELOPING A PROTOTYPE

SUPERMARINE TYPE 300 (K5054)
As flown by Joseph 'Mutt' Summers on its first flight, Eastleigh, 5 March 1936.

SUPERMARINE TYPE 300 (K5054)
As it appeared when flown by Flt. Lt. Humphrey Edwardes-Jones at the Aeroplane & Armaments Establishment, RAF Martlesham Heath, May 1936.

great rival, the Hawker Hurricane. The flight was uneventful, though a problem with the undercarriage uplocks meant that the wheels remained down for the duration of the first flight. The undercarriage was retracted in flight for the first time by Mutt Summers on 10 March, and after an engine change, Summers handed over flight testing to his assistants Jeffrey Quill and George Pickering. The aircraft was proving to be extremely promising, though the top speed was a disappointing 531km/h (330mph). An improved propeller increased that to 560km/h (348mph), after which K5054 was handed over to the RAF for service testing. The urgency attached to the Spitfire programme can be judged by the fact that RAF test pilot Flight-Lieutenant Humphrey Edwardes-Jones was asked to fly the aircraft and make his report to the Air Ministry immediately upon landing. After the flight, the pilot's only recommendation was that the aircraft be fitted with an undercarriage position indicator. A week later, the Air Ministry placed an order for 310 Spitfires.

TYPE 300

Weight (maximum take-off): 2359kg (5200lb)
Dimensions: Length 9.12m (29ft 11in), Wingspan 11.23m (36ft 10in), Height 2.51m (8ft 3in)
Powerplant: one 738kW (990hp) Rolls-Royce Merlin C liquid cooled V-12 piston engine
Maximum speed: 567km/h (349mph)
Range: 645km (400 miles)
Ceiling: 10790m (35,400ft)
Crew: 1

Mks I–III: Battle of Britain

With the production contract placed for the first batch of Spitfires, work proceeded on modifying K5054 to production standard, and several improvements to the airframe were developed and tested. Flight testing soon revealed that the rudder was overbalanced, and a reduction in the area of the aerodynamic balance yielded the familiar rudder shape that would grace the first few thousand production aircraft. The prototype had initially flown with a 990hp Rolls-Royce Merlin C – this was removed and replaced with a 1035hp Merlin F, which was representative of the production standard Merlin I. This engine would, in turn, be replaced by a Merlin II combined with Rolls-Royce-developed thrust-producing ejector exhausts which were calculated to add around 16km/h (10mph) to the top speed.

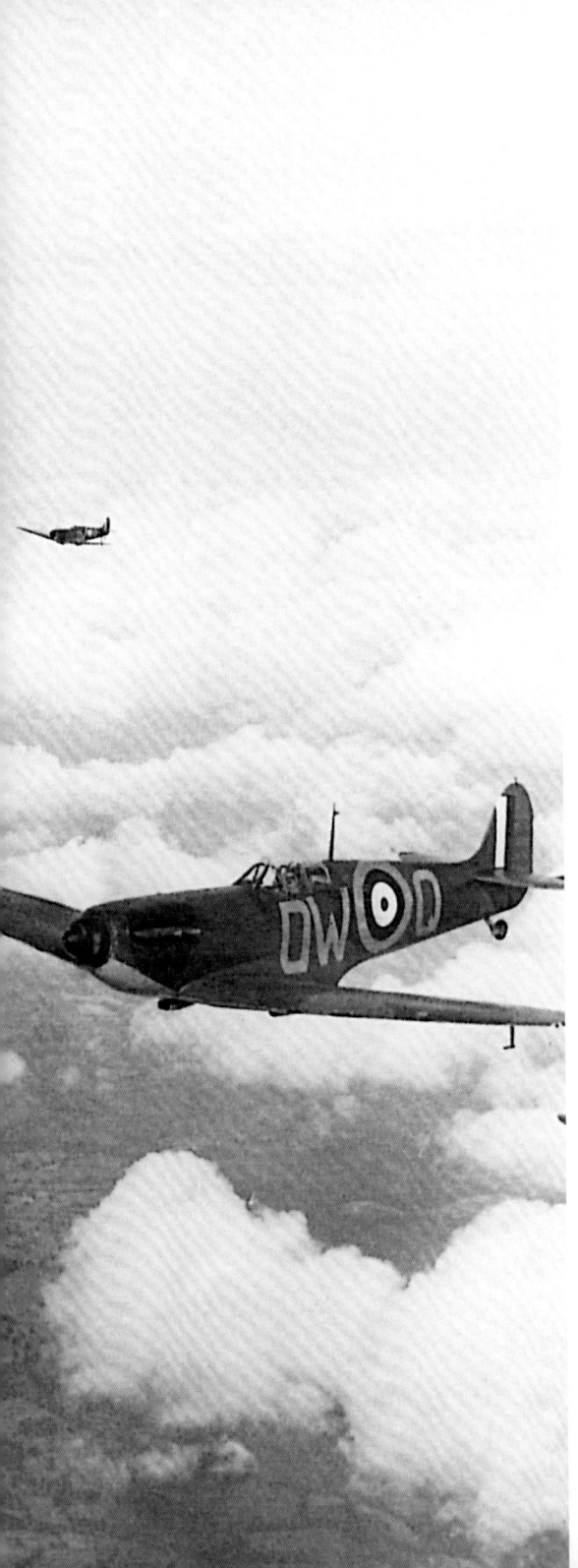

BATTLE FORMATION
From perhaps the most famous sequence of photographs depicting Spitfires in action during the Battle of Britain, this view shows aircraft of No. 610 Sqn, based at RAF Biggin Hill, on patrol during June 1940. Lessons learned in combat are reflected in the loose line-astern formations flown by the two sections in this view.

Mks I–III: Battle of Britain

TYPE 300 PROTOTYPE K5054, still in its all-over blue finish, was at Martlesham Heath in July 1938 for an inspection of the latest RAF aircraft types by King Edward VIII. RAF test pilots were enthusiastic in their praise for the new Supermarine fighter.

Other relatively minor changes to the aircraft saw the tailskid replaced with a tailwheel and the maximum flap deflection increased to 90 degrees to improve landing performance. A knock-out panel was also added to the cockpit canopy. This tiny change, though irrelevant to the performance of the aircraft, would mean the difference between life and death for many Spitfire pilots: the small Perspex panel could easily be pushed out by hand, the effect of which was to equalize the air pressure between the inside and outside of the canopy and make it possible to open the sliding hood at any speed and escape the aircraft in an emergency.

Camouflage finish

Befitting K5054's new role as a pre-production airframe, radio equipment and armament were fitted, and the aircraft received a camouflage finish. The armament of eight 7.7m (0.303in) Browning machine guns, made under licence by BSA and Vickers (Supermarine's parent company), had been stipulated after the basic design of the Spitfire, with four wing guns, was finalized. As a result, the four added guns were essentially shoehorned into the wing wherever they would fit, resulting in an irregular spanwise spacing between the weapons. The installation on the Hawker Hurricane that had been designed from the outset as an eight-gun fighter was far neater, and the arrangement of the weapons in the Spitfire's wing was to have unexpected effects on the combat performance of the fighter when it was committed to combat. K5054 continued flying

Mks I–III: BATTLE OF BRITAIN

on development work until it was written off in a landing accident at Farnborough on 4 September 1939, the day after the United Kingdom declared war on Germany. By this time, of course, production Spitfire Mk Is were serving in numbers with the RAF, the first example of a front-line unit having been delivered to 19 Squadron on 4 August 1938.

Joseph Smith
Sadly, RJ Mitchell would never witness his fighter enter service with the RAF. He had been diagnosed with rectal cancer in the summer of 1933, leading to emergency surgery in August of that year, and required the use of a permanent colostomy bag for the rest of his life. Despite the obviously debilitating effects of such a procedure, Mitchell was back at work in early 1934, and his staff never even knew of his condition. Amazingly, he also found time to acquire his pilot's licence during the same year. Unfortunately, the cancer returned in 1936, and Mitchell died on 11 June 1937 at the age of 42, unaware of the spectacular effect his design would have just a few short years later and the colossal success it would enjoy as a fighting machine. His place as chief designer at Supermarine was briefly taken up by Major Harold 'Agony' Payn but his tenure ended very soon after it began – he was not given security clearance for his new job by the Air Ministry because he was married to a German-born woman. As a result, it would be Joseph Smith,

CAMOUFLAGE
Modified to initial production standard during 1937, with radio aerial mast behind the cockpit and Merlin II engine with ejector exhausts fitted, K5054 models the dark earth and dark green scheme of service aircraft. Undersides were silver.

Mks I–III: BATTLE OF BRITAIN

Mks I–III: BATTLE OF BRITAIN

previously head of the drawing office, who, as chief designer, would steer the Spitfire through the manifold changes to the design that were to take place over the next decade or so. A highly experienced engineer, Smith has never received rightful credit for his realization that the Spitfire offered a near-limitless scope for development to remain in the vanguard of fighter design for the duration of the war.

Early production

Initial production of the Spitfire proceeded slowly. For example, by the end of 1937, a grand total of six fuselages had been produced at Supermarine and were awaiting wings from subcontractors. This was despite assurances that deliveries would begin in September and 60 would be delivered by the end of the year. The Air Ministry started to become extremely concerned about the delays as the deteriorating political situation in Europe meant that war was perceived to be ever more likely. The problem with Spitfire production was essentially threefold. Firstly, the aircraft was far more complex than anything Supermarine had built in quantity thus far; all the systems broke new

PREWAR SERVICE
These Spitfire Is of 65 Squadron pictured in early 1939 have already been fitted with the bulged cockpit canopies but retain the original aerial mast. The aircraft nearest the camera is being flown by future ace Robert Stanford-Tuck.

ground for the company, from the pneumatic equipment used for the brakes to the hydraulics for the retractable undercarriage. Additionally, the structure was complex and beyond Supermarine's ability to build in-house, so numerous subcontractors were relied upon. For example, Reynolds Tubes supplied the main spar, while the Pressed Steel Company fabricated the complex compound curves of the leading-edge skins.

Secondly, Supermarine was a fairly small company, employing a workforce of around 1300 in 1936 and still operating from the premises that Noel Pemberton-Billing had purchased in 1913. Somewhat counterintuitively, the small size of the Spitfire itself was also an issue. With the large flying boats Supermarine was accustomed to building, some 30 or 40 technicians could work on the aircraft at the same time. But in a Spitfire fuselage, only two or three workers could squeeze into the fuselage at a time.

Thirdly, Supermarine was very badly managed. There were no proper records kept regarding components and parts, and chaos reigned throughout the works. This situation was gradually turned around by management, who implemented proper processes while the factory was simultaneously modernized, but the early production of the Spitfire was described by test pilot Jeffrey Quill as 'traumatic'. So seriously did the Air Ministry take the situation – described by the Secretary of State for the Air Viscount Swinton as 'a disgraceful state of affairs' – that the Government demanded that Vickers restructure its aircraft division, and Sir Robert McLean, Chairman of Vickers Aviation, was replaced. Swinton himself would be forced to resign on 14 May 1938 because not a single Spitfire had yet been delivered.

Ironically, on that very day, the first production Spitfire I, K9787, flew for the first day of its flight trials in the skilled hands of Jeffrey Quill. Quill assumed this would be the first of a regular supply of production machines for him to flight test, but it would be another two months before the next one appeared. In a bid to solve the production problems, William Morris, Lord Nuffield, head of Morris Motors and the most famous practitioner of mass production in the UK, was asked by the Government to build a new factory to produce Spitfires and immediately awarded a contract for 1000 aircraft, the single biggest aircraft order made to date in Britain. Unfortunately, Nuffield's Castle Bromwich factory would also, at least initially, prove to be a massive disappointment.

First squadrons

Nonetheless, by the late summer of 1938, a trickle of production aircraft was beginning to reach RAF squadrons. Forty-five had been built by the end of the year, and both 19 Squadron and 66 Squadron possessed a full complement of Spitfires by December 1938. Early service use of the aircraft saw pilots repeatedly striking their heads on the inside of the canopy. This led to the adoption of the familiar humped canopy on production machines, and it was also retrospectively fitted to earlier aircraft, replacing the earlier flat-topped canopy. In the first half of 1939, a further six squadrons re-equipped with Spitfires, with two further units receiving Spitfires just as hostilities began during September. By this time, a modification to the aircraft that was to prove significant later in its development had already been undertaken: in June 1939, a trial installation of two 20mm (0.79in) Hispano cannon had been made, each fitted with a 60-round ammunition drum. Firing trials took place at Orford, on the east coast of Suffolk, but had not proved an overwhelming success as the weapons were prone to jamming.

Modifications to the ejector chute appeared to solve the problem, and 30 cannon-armed Spitfires were ordered and designated the Mk IB. Some of these Spitfires would see limited service with 19 Squadron in the Battle of Britain, but the cannon stoppage 'solution' turned out to be premature. Cannon failures proved prohibitively commonplace, and it would be some time yet before a cannon-armed Spitfire would be mass-produced.

First combat

The 'Phoney War' period would see the first tentative engagements

Mks I–III: BATTLE OF BRITAIN

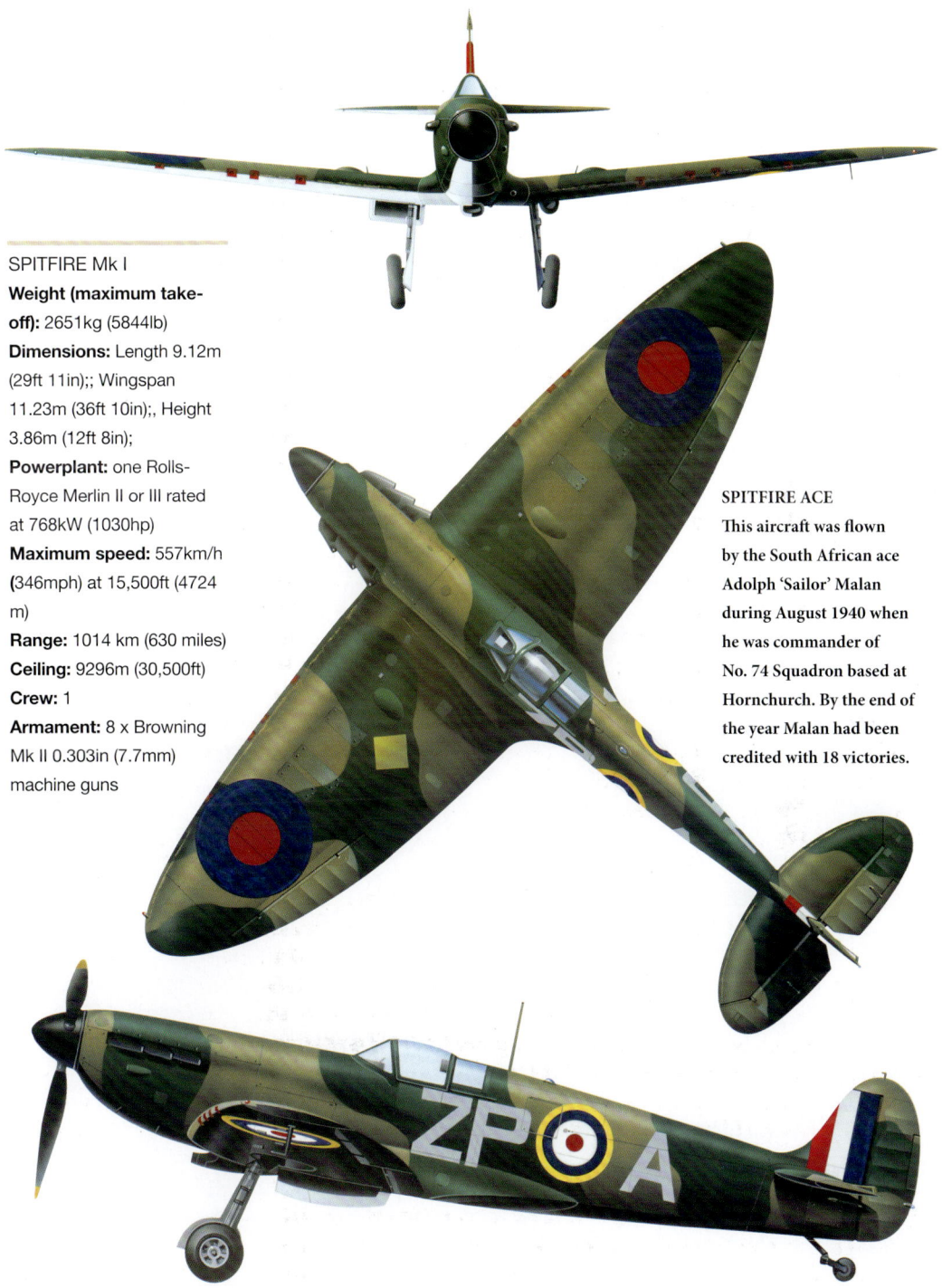

SPITFIRE Mk I
Weight (maximum take-off): 2651kg (5844lb)
Dimensions: Length 9.12m (29ft 11in);; Wingspan 11.23m (36ft 10in);, Height 3.86m (12ft 8in);
Powerplant: one Rolls-Royce Merlin II or III rated at 768kW (1030hp)
Maximum speed: 557km/h (346mph) at 15,500ft (4724 m)
Range: 1014 km (630 miles)
Ceiling: 9296m (30,500ft)
Crew: 1
Armament: 8 x Browning Mk II 0.303in (7.7mm) machine guns

SPITFIRE ACE
This aircraft was flown by the South African ace Adolph 'Sailor' Malan during August 1940 when he was commander of No. 74 Squadron based at Hornchurch. By the end of the year Malan had been credited with 18 victories.

HIGH SPEED SPITFIRE

An interesting one-off pre-war development occurred around the time the first Mk Is were being delivered to squadrons in the form of the High Speed Spitfire. Supermarine had been considering attempting the World Landplane Speed Record as early as August 1937. The Spitfire was the fastest fighter aircraft in the world when it first appeared, and it was believed that a modified aircraft could be pushed up to a speed of around 644km/h (400mph). The programme centred around a racing version of the Rolls-Royce Merlin, which was estimated to be capable of delivering around 2000hp fitted to a standard Mk I airframe. This aircraft was given the manufacturer's registration N17 and was finished in a highly polished blue and silver finish. A new low-drag cockpit canopy and windscreen were fitted, and the tailwheel was replaced with a skid, whilst the wing received blunt wingtips that reduced the span. The racing Merlin required a larger radiator to be fitted to deliver adequate cooling. This resulted in a considerable increase in drag, but even with this impediment the High Speed Spitfire attained a speed of 657km/h (408mph) in the spring of 1939.

A different cooling system was proposed, dispensing with the radiator altogether and allowing the coolant to boil in a special tank in the wing. The Royal Aircraft Establishment (RAE) estimated that the aircraft would be capable of 684km/h (425mph) in this form. Unfortunately for the High Speed Spitfire, the Messerschmitt Me 209 V1 set a record of 755.14km/h (469.22mph) in April 1939, which would officially stand until November 1945. This speed was beyond the capability of the Spitfire to attain, and no record attempt was ever made. The High Speed Spitfire was later fitted with a standard Merlin XII engine and was used as a 'hack' by the Photographic Reconnaissance Unit (PRU). In this form it would survive the war and was finally struck off charge in August 1946 and scrapped.

HIGH SPEED SPITFIRE
Resplendent in its royal blue and silver colour scheme, the High Speed Spitfire was fast but not fast enough to gain the absolute airspeed record. Note the enlarged underwing radiator, blunt wingtip and low drag canopy.

between Spitfires and German aircraft. However, the Spitfire's first 'kill' was, tragically, a friendly-fire incident, later known as the 'Battle of Barking Creek'. On 6 September 1939, a mere three days after the British declaration of war, a radar fault meant that several unidentified aircraft appeared to be approaching from the east, and four squadrons scrambled to meet the perceived threat. Few RAF aircraft had yet been fitted with IFF (Identification Friend or Foe) sets and none of the pilots in the sky had encountered a German aircraft before. On edge and expecting to meet enemy aircraft for the first time, the Spitfire pilots of 74 Squadron misidentified the Hurricanes of

SPITFIRE MK I

The first squadron to equip with the Spitfire was 19 Squadron based at Duxford in Cambridgeshire. K9797 was the 11th production Spitfire and was written off in May 1939 after an engine fire and forced landing by future ace, George 'Grumpy' Unwin.

151 Squadron as enemy aircraft and attacked. With remarkable presence of mind, 151 Squadron Leader Edward Donaldson realized the attackers were British and ordered his squadron not to retaliate. Meanwhile, 74 Squadron rapidly realized their error and broke off but not before two Hurricanes had been shot down with one pilot losing his life. The same action would also result in the first combat loss of a Spitfire when a British anti-aircraft battery managed to shoot down one of the fighters as they returned to base. Despite the dismal outcome of the day's unfortunate events, the Battle of Barking Creek did result in the exposure of the inadequacy of RAF radar and identification procedures, leading to both being greatly improved by the time of the Battle of Britain. Thus, the battle may have saved more lives and aircraft in the long run than were lost at the time.

The Spitfire would not have to wait long though before it would meet genuine enemy aircraft. On 16 October 1939, six Spitfires intercepted nine Junkers Ju 88s of 1./KG30 over Rosyth, Scotland, whilst they attempted to attack two Royal Navy cruisers in the Firth of Forth. Two of the attacking aircraft were shot down and a third was heavily damaged. Similar sporadic actions against German bombers operating singly or in small groups would form the bulk of early Spitfire operations even as the Battle for France escalated. Air Chief Marshal Hugh Dowding, head of Fighter Command, realized that the Spitfire was a more effective defensive fighter than the Hurricane and resisted calls for Spitfire squadrons to be sent to France, instead husbanding his precious Spitfire force within the British Isles. The Dunkirk evacuation of May 1940, however, was of sufficient importance that even the Spitfires were

SPITFIRE Mk I

Weight (maximum take-off): 2651kg (5844lb)
Dimensions: Length 9.12m (29ft 11in), Wingspan 11.23m (36ft 10in), Height 3.86m (12ft 8in)
Powerplant: one Rolls-Royce Merlin II or III rated at 768kW (1030hp)
Maximum speed: 557km/h (346mph) at 15,500ft (4724 m)
Range: 1014 km (630 miles)
Ceiling: 9296m (30,500ft)
Crew: 1
Armament: 8 x Browning Mk II 0.303in (7.7mm) machine guns

committed, along with virtually every other available fighter. This operation would constitute the first large-scale action in which the Spitfire was involved.

Dunkirk's proximity to the British coast meant that even an aircraft designed as a short-range interceptor could operate for a meaningful period over the beaches, and although the fighters had to operate without the benefit of radar cover, 132 Luftwaffe aircraft were destroyed during the Dunkirk evacuation. Unfortunately, this was at the cost of 99 fighters lost, 38 of them Spitfires.

New propellor
By this time the Spitfire's original two-blade wooden propeller had been replaced by a three-blade de Havilland metal propeller with two pitch positions: 'fine' for take-off and 'coarse' for high-speed flight. Tests by the Aircraft and Armament Experimental Establishment (A&AEE) in 1939 had shown that the new propeller conferred a slightly greater maximum speed on the Spitfire, albeit at the expense of climb performance. The wooden airscrew was also considerably lighter than the metal unit, requiring 61kg (134.48lb) of ballast to be carried in the nose when fitted. A further improvement came with the fitting of constant-speed propellers – the result of an urgent modification programme in late June and July 1940 to equip all operational Spitfires and Hurricanes with

SUPREME INTERCEPTOR
A Heinkel He 111 erupts in smoke in this, somewhat touched-up, still from Adolph 'Sailor' Malan's gun camera. The undercarriage has lowered due to damage to the hydraulic system.

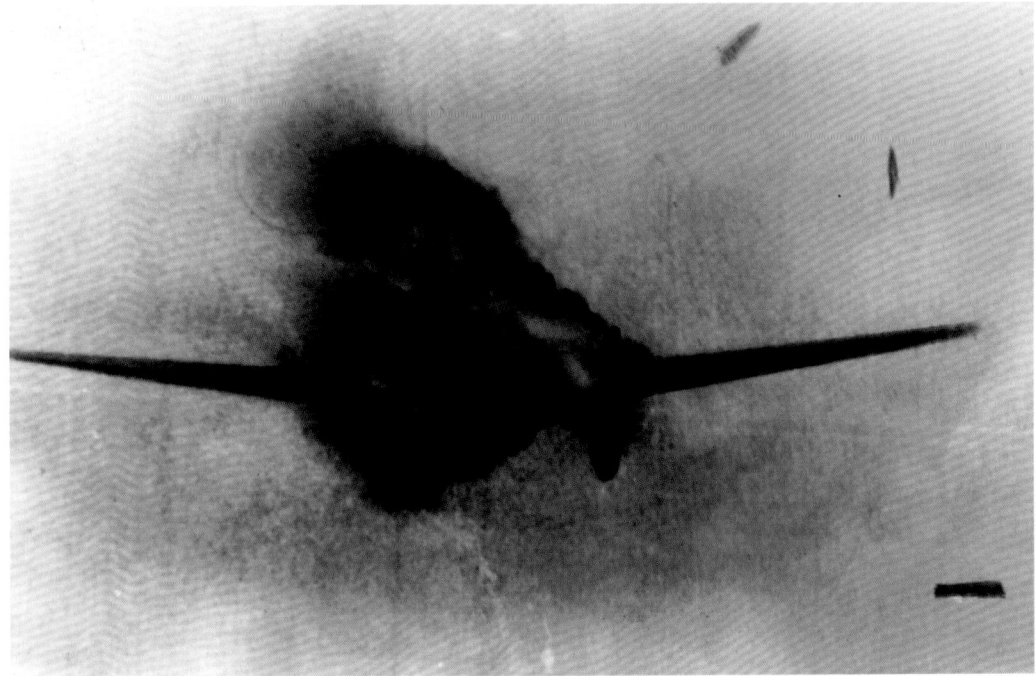

Mks I–III: BATTLE OF BRITAIN

constant-speed de Havilland airscrews. This action had been prompted by the acquisition of an intact Messerschmitt Bf 109 in May 1940. In tests against the Hurricane and Spitfire, the 109 had proved to possess superior performance to both when fitted with the DH two-position airscrew. However, when a constant-speed unit was fitted, the Spitfire broadly matched the 109 for speed and climb. The conversion process was managed by de Havilland with a remarkable efficiency hitherto unusual in the Spitfire story: a team of engineers working up to 15 hours a day succeeded in converting 1054 Spitfires and Hurricanes within a month.

SPITFIRE MK IA
P9374 was built in February 1940 and served with 92 Squadron over Dunkirk. After force landing on a beach near Calais, P9374 was covered by the shifting sands only to re-emerge in 1980. Following a long restoration, P9374 flew again in 2011.

Battle of Britain
The fall of France meant that the stage was set for the Spitfire's most famous single action: the Battle of Britain. Despite this being the event with which Spitfires are most associated, they were not the most numerous of the RAF fighters involved in the battle: roughly two-thirds of operational fighters were Hawker Hurricanes. Hurricanes were responsible for the majority of losses inflicted on the Luftwaffe, famously being responsible for more aircraft shot down than all other defences combined. Yet the Spitfires shot down more enemy aircraft relative to their own numbers: 33 Hurricane squadrons shot down 656 aircraft, while the 18 Spitfire squadrons managed to account for 529 victories – a considerably better result given the relative fleet sizes.

In comparison to its great partner and rival, the Hurricane, which used exactly the same engine, the Spitfire was faster and had a greater rate of climb

---

SPITFIRE Mk IA
Weight (maximum take-off): 2651kg (5844lb)
Dimensions: Length 9.12m (29ft 11in); Wingspan 11.23m (36ft 10in); Height 3.86m (12ft 8in);
Powerplant: one Rolls-Royce Merlin II or III rated at 768kW (1030hp)
Maximum speed: 557km/h (346mph) at 15,500ft (4724 m)
Range: 1014 km (630 miles)
Ceiling: 9296m (30,500ft)
Crew: 1
Armament: 8 x Browning Mk II 0.303in (7.7mm) machine guns

Mks I–III: BATTLE OF BRITAIN

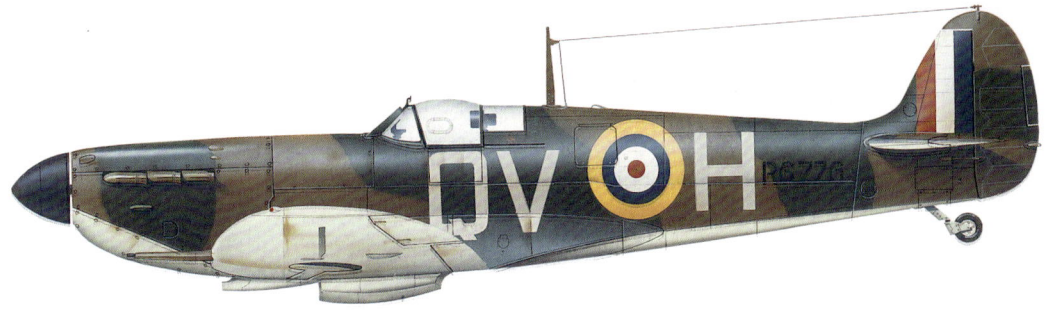

SPITFIRE Mk IB
Weight (maximum take-off): 2431kg (5339lb)
Dimensions: Length 9.12m (29ft 11in), Wingspan 11.23m (36ft 10in), Height 3.02m (9ft 10in)
Powerplant: 770kW (1030hp) Rolls-Royce Merlin III liquid cooled V-12 piston engine
Maximum speed: 580km/h (365mph)
Range: 645km (400 miles)
Ceiling: 10,485m (34,400ft)
Crew: 1
Armament: two 20mm (0.79in) Hispano cannon and four 7.7mm (0.303in) Browning machine guns in wings

SPITFIRE MK IB
One of the few Mk Is to be equipped with 20mm cannon, R6776 served with 19 Squadron at Duxford throughout the Battle of Britain and was regularly flown by six-kill ace Brian Lane.

thanks to its more advanced aerodynamics. A rather convoluted RAF test in July 1939 saw 12 Hurricanes flying 274.32m (900ft) in front of 12 Spitfires at 5486.4m (18,000ft). At a given signal, the pilots all opened up to full throttle and the individual speeds of all the aircraft involved were recorded when the Spitfires overtook the Hurricanes. On average, the Spitfires were flying 68km/h (42mph) faster than the Hurricanes at the point they overtook.

On the other hand, the Hurricane's machine guns were grouped together in a better-designed fitting than the Spitfire, and it was justly famed as an excellent 'gun platform'. By contrast, the recoil from its eight Brownings could reduce the Spitfire's speed by 40km/h (25mph) and cause the nose to pitch down much as if the flaps were deployed for landing. The outboard guns of the Spitfire were fitted in an area prone to flexing, firing the guns caused the wings to judder, and its accuracy of fire was never as good as that of the Hurricane.

Spitfire compared
The other advantage that the Hurricane appeared to possess was its toughness. The Hawker fighter's airframe was immensely strong and its traditional construction of fabric over a wood and metal frame meant that cannon shells would often pass straight through the fuselage, leaving no damage worse than a small hole that could be easily patched. By contrast, the Spitfire was built with a stressed skin wherein the aluminium covering of the airframe is a load-bearing part of the structure. A cannon shell would cause far greater damage when it hit the skin and could compromise the structural integrity of the aircraft. Combat damage inflicted on stressed-skin panels is also more difficult to repair. Conversely, the Hurricane possessed a terrifying shortcoming not shared by the Spitfire – a fuel tank was mounted in each wing at the root, right

MESSERSCHMITT BF 109 ADVERSARY

The only German single-engine fighter to fly in the Battle of Britain was the Messerschmitt Bf 109, and it was widely considered by the Germans to be the finest fighter aircraft in the world – and with some justification. Unlike the Spitfire, the Bf 109 was thoroughly combat-proven, having seen service in the Spanish Civil War since mid-1937 before being unleashed against Poland and France, and it had proved superior to all other fighter aircraft it had encountered. Compared to the Spitfire Mk I, the Bf 109E was smaller and lighter but possessed higher wing loading. Maximum speed and climb rate were very evenly matched but as previously mentioned, the Spitfire could turn tighter than the Messerschmitt.

However, the 109 had two critical advantages over the British fighters: its armament and its fuel-injected engine. Except for the cannon-armed Spitfire Mk IBs, which proved unreliable, all Spitfires and Hurricanes involved in the Battle of Britain were armed with eight rifle-calibre machine guns, this having been considered highly powerful when the aircraft were schemed back in the mid-1930s. By contrast, the 109 was fitted with two calibres of weapon. In the 109E-4, which was the most commonplace variant in the Battle of Britain, this consisted of two rifle calibre 7.92mm (0.31in) MG 17s in the upper engine cowling and one 20mm (0.79in) MG FF/M cannon mounted in each wing. The cannon packed a much greater punch than the machine gun, and any hits scored on enemy aircraft were of much greater destructive power. On the other hand, the ammunition supply of the 109 was extremely limited, allowing for only seven seconds of firing time from the cannon compared to around 17 seconds for the machine guns of the Spitfire. Nonetheless, the fact that both Spitfire and Hurricane would later be armed with 20mm (0.79in) cannon as standard would suggest that larger calibre was more important than total ammunition supply.

The other huge advantage the Bf 109 enjoyed was that its engine was fed by fuel injection rather than the conventional float carburettor of the Merlin. This meant that the 109's fuel feed was unaffected by attitude or G-load. If being pursued by a British fighter, a 109 could simply nose over into a dive: if the pursuing aircraft attempted to follow, the negative G would cause the fuel in the carburettor float chamber to rise to the top, temporarily starving the engine of fuel and allowing the Messerschmitt to escape. If negative G continued, the float chamber would fill with fuel, forcing the float to the bottom. Since the float controlled the needle valve regulating fuel intake, the carburettor would flood, drowning the supercharger with an over-rich mixture. The consequent rich mixture cut-out would then cause the engine to shut down. The problem could be ameliorated by rolling the aircraft inverted as the dive was entered to maintain a positive G-load and fuel flow, but this inevitably slowed down the aircraft, and throughout the Battle of Britain, pilots made impassioned requests to solve this issue.

Interestingly, Rolls-Royce had considered using fuel injection during the early stages of Merlin development but chose the carburettor because early mechanical fuel injection systems were complex and unreliable and because in a carburettor-fed engine, fuel vaporizing in the inlet cooled the charge, increasing power output. Anything that sapped power and therefore speed and rate of climb was to be avoided, particularly as the Spitfire was expected to be dealing more with unescorted German bombers than with manoeuvrable fighters. The problem was partially solved by a simple brass thimble-shaped device with a small hole that possessed precisely calculated dimensions to allow just enough fuel flow for maximum engine power. This temporarily prevented the float chamber from flooding with fuel and was sufficient to allow for brief negative G manoeuvres. Invented by the brilliant aeronautical engineer Beatrice Shilling at the Royal Aircraft Establishment, the device was officially called the R.A.E. Restrictor but quickly gained the puerile nickname 'Miss Shilling's Orifice'.

next to the cockpit. If hit by gunfire these tended to catch fire immediately, leading to the temperature in the cockpit rising to 3000°C (5432°F) within 10 seconds. Analysis of combat following the Battle of Britain suggested that contrary to popular belief, the Spitfire was the more resilient machine in combat.

Even at the time, the relative manoeuvrability of the Spitfire was hotly debated. Some pilots stated outright that the Hurricane could out-turn the Spitfire, whilst others claimed the reverse. Both could out-turn the Bf 109, though even here there is some doubt: in perfect conditions, the 109 could theoretically beat both the RAF machines in a turn. The reality of combat revealed that in practice the opposite was the case, and it would be a foolish Luftwaffe pilot who attempted to take on either British machine in turning combat. In action, the Spitfire's higher performance saw it being utilized, wherever possible, to deal with the escort fighters while the Hurricanes attacked the bombers. The reality of combat, however, seldom allowed for such a tidy separation of tasks, and at least 179 Bf 109s are known to have been lost to Hurricanes.

EAST INDIA SQUADRON
92 Squadron converted from Bristol Blenheims to Spitfires from March 1940 and flew various marks of the Supermarine fighter until re-equipping with the Gloster Meteor in 1947. Here Squadron Leader John Kent chalks up the squadron's 130th victory of the war.

Mks I–III: BATTLE OF BRITAIN

SPITFIRE Mk II
Weight (maximum take-off): 2864kg (6275lb)
Dimensions: Length 9.12m (29ft 11in), Wingspan 11.23m (36ft 10in), Height 3.02m (9ft 10in)
Powerplant: one 846kW (1135hp) Rolls-Royce Merlin XII liquid cooled V-12 piston engine
Maximum speed: 595km/h (370mph)
Range: 645km (400 miles)
Ceiling: 11,460m (37,600ft)
Crew: 1
Armament: eight 7.7mm (0.303in) Browning machine guns in wings

Top: **SPITFIRE MK II**
The grey and white colour scheme of this Mk II of the Czech-manned 312 Squadron, was adopted by RAF Fighter Command to better camouflage the aircraft over water, reflecting the offensive sweeps over occupied France regularly conducted after the Battle of Britain.

There was also a third British single-engine monoplane fighter that saw service during the Battle of Britain: the Boulton Paul Defiant. An unusual fighter, the Defiant possessed no fixed forward-firing armament, instead being equipped with a manned powered turret armed with four machine guns and thus able to shoot in virtually any direction. For

Lower: **SPITFIRE MK IIA**
Another Spitfire flown by non-British Commonwealth personnel, this Mk II was on the strength of 340 Squadron, a Free French unit, as denoted by the Cross of Lorraine on the nose.

a period just before the war, the Air Ministry was extremely keen on the 'turret fighter' concept, and many were convinced that it was likely to supersede conventional fighters. No less a personage than Winston Churchill was one such, writing in 1938 that 'we should now build, as quickly and in as large numbers as we can, heavily armed aeroplanes designed with turrets' adding that 'the urgency for action arises from the fact the Germans must know we have banked upon the forward-shooting plunging "Spitfire" whose attack must most likely resolve itself into a pursuit which, if not instantly effective, exposes the pursuer to immediate destruction'. The Defiant was an excellent aircraft, first flown without the turret and demonstrating performance somewhere between the Hurricane and Spitfire. The addition of the heavy turret and a crewmember to operate it had an inevitably deleterious effect on its performance, however.

The Defiant, whilst briefly enjoying great success against unescorted bombers, was unable to cope with the Messerschmitt 109, although on one occasion a group of 109s mistook the Defiants for Hurricanes and dived upon them from the rear only to be met with withering fire from the turrets.

However, both Defiant squadrons committed to the Battle of Britain suffered tremendous losses at the hands of enemy fighters and were withdrawn from daylight operations. Churchill's great hope had proved unfortunately misguided.

Mk II

The first of many significant changes to the powerplant of the Spitfire appeared in 1940 when a new Merlin variant, the XII, became available. This was the first production Merlin to utilize 100 octane fuel as opposed to the 85 octane fuel in use with the Merlin II and III as fitted to Spitfire Mk Is, and was rated at 1175hp, an increase of over 100hp over the earlier engine. The new engine resulted in the aircraft being designated the Spitfire Mk II. It also featured increased armour production for the pilot and an improved cooling system but was externally identical to the Mk I save for a tiny fairing over the Coffman starter behind the propeller spinner on the starboard side. Spitfire IIs started to leave the production line in June 1940 with the first squadrons receiving the aircraft in August.

All Mk II Spitfires were built at the Nuffield Castle Bromwich Aircraft Factory – 751 with the regular eight-gun armament, later designated the Mk IIA, as well as 170 Mk IIBs with two 20mm (0.79in) cannon and four machine guns, a new feed and cartridge ejection system having

---

SPITFIRE Mk IIA
Weight (maximum take-off): 2651kg (5844lb)
Dimensions: Length 9.12m (29ft 11in), Wingspan 11.23m (36ft 10in), Height 3.86m (12ft 8in)
Powerplant: one Rolls-Royce Merlin II or III rated at 768kW (1030hp)
Maximum speed: 557km/h (346mph) at 15,500ft (4724 m)
Range: 1014 km (630 miles)
Ceiling: 9296m (30,500ft)
Crew: 1
Armament: 8 x Browning Mk II 0.303in (7.7mm) machine guns

SPITFIRE MK IIA
P7666 of 41 Squadron, named 'Observer Corps' was regularly flown by Squadron Leader Donald Finlay, Gold medallist in the 110m hurdles at the 1936 Berlin Olympics. P7666 was lost in combat with Bf 109s in April 1941, with pilot Jack Stokoe parachuting to safety.

Mks I–III: BATTLE OF BRITAIN

SPITFIRE MK IIB
This cannon-armed presentation Spitfire, named 'Ceram', was served with the Polish-manned 306 Squadron based at Northolt. Sergeant Marcin Machowiak claimed a Bf 109 shot down in this aircraft on 29 August 1942.

SPITFIRE Mk IIB
Weight (maximum take-off): 2846kg (6275lb)
Dimensions: Length 9.12m (29ft 11in), Wingspan 11.23m (36ft 10in), Height 3.02m (9ft 10in)
Powerplant: one 846kW (1135hp) Rolls-Royce Merlin XII liquid cooled V-12 piston engine
Maximum speed: 595km/h (370mph)
Range: 645km (400 miles)
Ceiling: 11,460m (37,600ft)
Crew: 1
Armament: two 20mm (0.79in) Hispano cannon and four 7.7mm (0.303in) Browning machine guns

been developed to overcome the problems experienced by the initial cannon-armed aircraft. Initial production from Lord Nuffield's new factory was painfully slow, largely as a result of poor management and inexperience – Nuffield was well versed in mass-producing cars but not aircraft.

Castle Bromwich was supposed to have produced 1000 Spitfires by the time of the Battle of Britain but had actually built none. Had those extra aircraft been available before the summer of 1940, the outcome of the battle would never have been in doubt.

Eventually, Lord Beaverbrook's Ministry of Aircraft Production removed Nuffield and ordered staff from Vickers to address the shortcomings at the plant. Within months, the situation was turned around, and the plant was ultimately responsible for more than half of total Spitfire production.

Mk III
With the Spitfire II, an aircraft that featured relatively minor improvements to the Mk I airframe, established in production, attention turned to an aircraft Supermarine had been working on since 1939 with the aim of producing a much more extensively modified and improved version. This aircraft featured a strengthened fuselage structure, an internal bulletproof windscreen, a retractable tailwheel, a strengthened main undercarriage with main wheels raked forward 5.08cm (2in) for improved ground handling, an enlarged underwing oil cooler and an additional 40kg (88.18lb) of armour.

The most obvious change, however, was to the wing, which featured bluntly rounded wingtips to reduce span-loading and improve rate of roll. The engine fitted was the Merlin R.M.3SM, rated at 1390hp and equipped with a two-speed supercharger, which would enter production as the Merlin XX. The new aircraft was capable of 644km/h (400mph) at 6400m (21,000ft) and an order for 1000 aircraft (designated Spitfire Mk III) was placed with the Castle Bromwich Aircraft Factory at the end of 1940. However, not a single production example would be built, for reasons that will become clear in the next chapter.

ERGONOMIC NIGHTMARE
Early Spitfire cockpits were not noted for their user-friendliness, with controls and instruments distributed about the cockpit anywhere they would fit. On take-off, the Spitfire pilot had to change hands from throttle to control column and back again to raise the undercarriage, which led to a distinctive undulating climb.

SCRAMBLE!
A classic image, as the French pilots of 340 Squadron sprint to their waiting Spitfire IIAs ready to face the enemy. This is a posed shot, in reality the aircraft would have been started by their ground crews ready for immediate take-off.

Mks V–VI: Spitfires of the World

With the Battle of Britain having concluded in its favour, the RAF began to pursue a more aggressive policy of carrying the fight to the enemy on the European mainland, a period known as 'leaning into France'. As a result, efforts began to focus on improving the Spitfire's range, never previously a huge concern for an aircraft intended as a home-defence interceptor. The first range-extending modification came in the form of a large 136.38L (30 gallon) flush-fitting non-jettisonable external tank that fitted flush under the port wing only and resulted in a distinctly lopsided appearance. One hundred kits of parts were produced after the tank was tested in September 1940 to be fitted when required.

DESERT PATROL
Two 601 Sqn LF Mk Vb Spitfires are led by Wing Commander Ian Gleed cruise over De Djerba Island, en route to the Mareth Line in North Africa in 1943. All three aircraft feature 'clipped' wingtips befitting their low-level role.

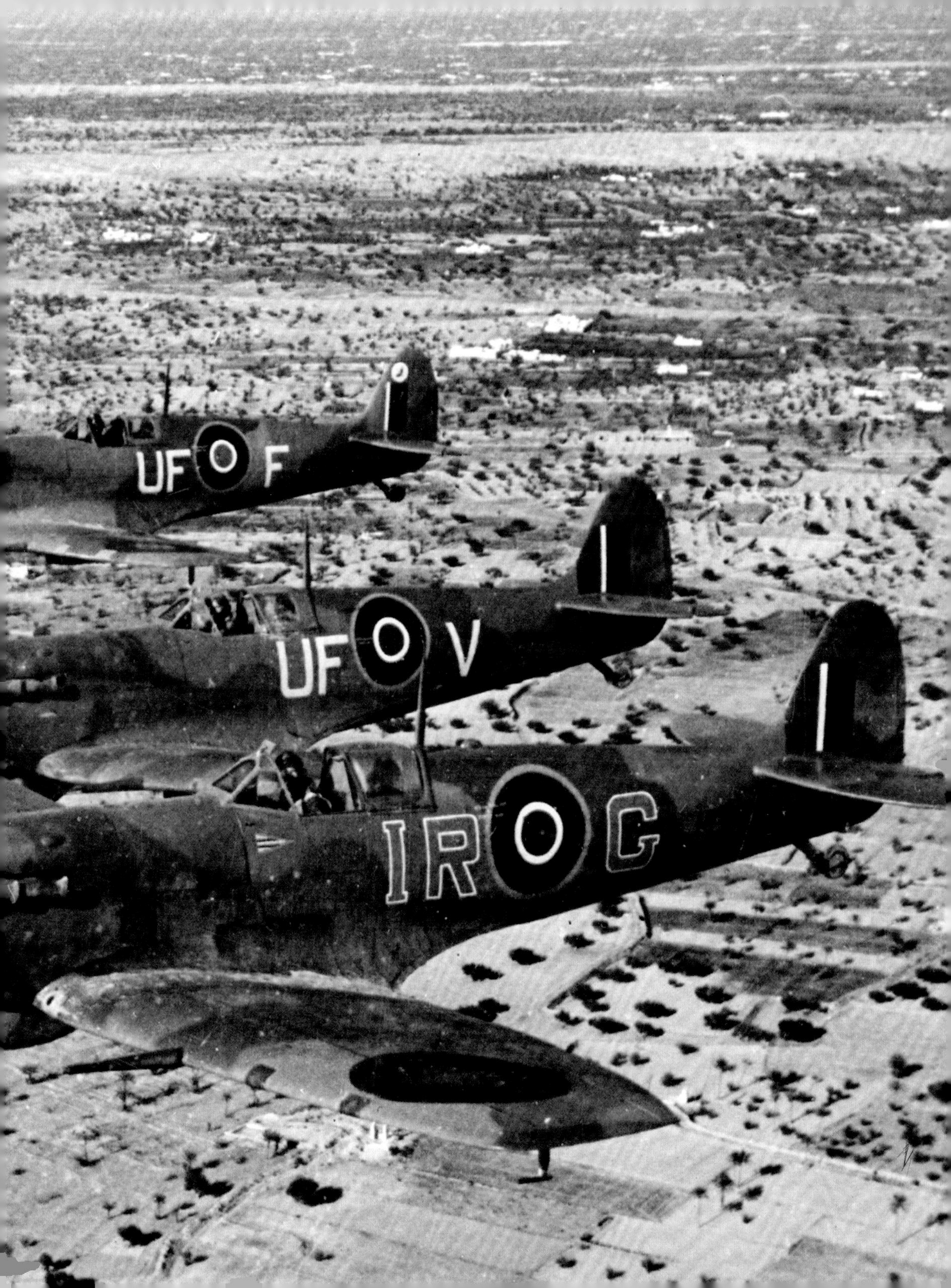
UF F
UF V
IR G

FIRING UP

Flames belch from the exhaust of this 222 Squadron Spitfire VB at North Weald in 1942 as the Merlin starts up. 222 Squadron flew Spitfires until December 1944.

In June 1941, five squadrons of Spitfires were required to provide escort for Operation Sunrise, a daylight attack scheduled to take place on 24 July 1941. The attack was focused on the Scharnhorst, Prinz Eugen and Gneisenau in Brest Harbour. This was the perfect opportunity to fit the tanks, which were used to modify around 50 Spitfire IIAs, these subsequently being designated Spitfire IIA (LR). Some of the long Spitfire IIA (LR)s were passed on to 19 Squadron for further long-range missions over the continent. However, they were unpopular due to their asymmetric trim, and the tanks were removed.

New Merlin engine

At much the same time as the first long-range tanks were being developed, Rolls-Royce began producing a new Merlin engine, the Merlin 45, featuring a new design of single-stage single-speed supercharger that delivered a combat rating of 1470hp at 2819.4m (9250ft). Tested in a Mk IA airframe, the Merlin 45-powered Spitfire entered flight trials in December 1940. It quickly became clear that this relatively simple conversion offered a similar level of performance improvement as the Mk III described in the previous chapter, only much more easily and rapidly. As a result, this aircraft entered production as the Mk V – the Mk IV being the prototype Griffon-powered Spitfire which would lead to a completely new line of development as described later. Existing production contracts for the Mk III were altered to cover the Mk V instead. The initial production Mk V was almost indistinguishable from earlier variants, the most obvious point of identification being the circular underwing oil cooler design adopted by the new aircraft, which had been designed for the Mk III. The lower nose contours and coolant radiator were also marginally different, but in most respects the aircraft appeared identical to its forebears. This was the first time that a carefully designed Spitfire variant that was intended to be the definitive version available was to be sidelined by a less sophisticated but more immediately available re-engined update of an earlier mark. The same process would occur on two further occasions during the Spitfire's production life.

Mks V–VI: SPITFIRES OF THE WORLD

Mk V: the greatest production model

The Mk V was destined to become the most produced single Spitfire variant, with 6479 built in a dizzying array of subtypes. It was also the first Spitfire variant to see service with the RAF overseas and the first to serve in the air forces of other nations. The initial 154 Mk Vs were conversions of existing Mk I and II airframes, undertaken

SPITFIRE MK VB

On the strength of 40 Squadron of the South African Air Force, as denoted by the orange roundel centres. This example features an oblique camera mounting behind the cockpit for the tactical reconnaissance role over Italy during 1943.

by Rolls-Royce, Supermarine and various other companies, and 92 Squadron became the first operational squadron to receive the new variant in May 1941. Production of new-build Mk Vs began at the CBAF in June 1941. Becoming widely available from mid-1941 onwards, the Mk V would become the type most heavily committed to the RAF's lean into France, with some 59 squadrons utilizing the Mk V for fighter sweeps over occupied Europe between 1941 and 1943.

'Slipper' tank

The ongoing issue of short-range was eased somewhat by the introduction of a 'slipper' tank that fitted flush to the belly of the

SPITFIRE Mk Vb
Weight (maximum take-off): 3071kg (6525lb)
Dimensions: Length 9.12m (29ft 11in), Wingspan (clipped) 10.01m (32ft 10in), Height 3.02m (9ft 10in)
Powerplant: one 1181kW (1585hp) Rolls-Royce Merlin 50M liquid cooled V-12 piston engine
Maximum speed: 564km/h (351mph)
Range: 756km (470 miles)
Ceiling: 10,881m (35,700ft)
Crew: 1
Armament: two 20mm (0.79in) Hispano cannon and four 7.7mm (0.303in) Browning machine guns in wings; up to 230kg (500lb) bomb load

SPITFIRE MK VB

The aircraft in the foreground of the photo on P32/33, AB502 is fitted with an 'Aboukir' dust filter over the carburettor intake and is coded with Wing Commander Gleed's initials 'IR G', a privilege of his rank.

CANNON ARMAMENT

The Mk V was also the first Spitfire variant for which cannon armament became the norm: only 94 Mk Vas with the previously standard eight machine gun armament were produced, as compared to 3911 Mk Vbs with one 20mm (0.79in) cannon replacing the two inboard 7.7mm (0.303in) Brownings in each wing. Later production would standardize on the Mk Vc with the C-type or 'Universal' wing.

As well as being redesigned to make it simpler and quicker to build, the C-type wing was engineered so that either eight machine guns, two cannon and four machine guns, or four cannon could be fitted. The four-cannon set-up, however, was seldom employed due to the weight of the cannons and ammunition and its deleterious effect on handling – aircraft fitted with four cannon at the factory often had one pair removed in the field.

The vast majority of Spitfire Vcs utilized the two cannon and four machine guns fit, and the unused cannon ports protruding from the leading edge were sealed with a rubber plug. At much the same time as the universal wing was appearing, a tropical version of the Spitfire was introduced with a large dust filter over the carburettor intake, intended for operations in the Middle East and Africa.

NEW WING, NEW ARMAMENT
From any angle, the Spitfire's elliptical wing was a masterpiece of engineering and aesthetics. The basic wing design was retained until work began on the Mk 21 wing in 1942. The two 20mm (0.79in) cannons are prominent from this angle.

aircraft and could be jettisoned in flight. Around 300,000 slipper tanks were manufactured for use with the Spitfire in a variety of capacities. Some difficulty had been experienced in early jettison trials with tanks damaging the rear fuselage and tailplane when released, but this had been solved through the simple expedient of fitting two stops a short distance behind the tank. On release, the tank would slide backwards until it hit the stops, whereupon the nose would tip forward and

Mks V-VI: SPITFIRES OF THE WORLD

ISLAND DEFENDER
Squadron Leader JJ Lynch, an American, commanded 249 Squadron on Malta and shot down a Junkers Ju 52/3m on 28 April 1943 in this Spitfire Mk Vc. This was calculated to be the 1000th victory for Malta-based aircraft, and Lynch's seventh.

the tank would fall away from the aircraft. The first tanks were of 136.38L (30 gallon) capacity, followed by a 204.57L (45 gallon) version, with a large 409.15L (90 gallon) tank available for use on special operations. One such operation was the remarkable mission in March 1942, which saw Spitfire Vcs flying off carriers in the Mediterranean to reinforce the fighter defences of Malta. This represented the first instance of Spitfires being deployed beyond the British Isles. Three groups of Spitfire Vbs, totalling 31 aircraft, fitted with large tropical dust filters and belly tanks, flew off the carriers *Eagle* and *Argus* sailing off the Algerian coast.

Further such missions followed, with 46 Spitfire Vcs flying off the American carrier *Wasp* on 20 April and 60 further aircraft from *Wasp* and *Eagle* on 9 May. Similar missions occurred at a regular pace until, by October, 367 of 385 Spitfire Vs launched had arrived in Malta.

Malta mission
Meanwhile, an even larger belly tank had been developed in a bid to endow the aircraft with sufficient range to make a direct flight from Gibraltar to Malta: a distance of 1770km (1100 miles). The new tank was huge, with a capacity of 772.84L (170 gallons), and completely enclosed the oil cooler intake, a duct through the tank itself maintaining the air

53

Mks V–VI: SPITFIRES OF THE WORLD

SPITFIRE Mk V
Weight (maximum take-off): 3078kg (6785lb)
Dimensions: Length 9.11m (29ft 11in); Wingspan 11.23m (36ft 10in); Height 3.48m (11ft 5in)
Powerplant: 1074kW (1440hp) Rolls-Royce Merlin 45/46/50 V-12 engine
Maximum speed: 602km/h (374mph)
Range: 756km (470 miles)
Ceiling: 11,280m (37,000ft)
Crew: 1
Armament: two 20mm (0.79in) cannon and four 7.7mm (0.303in) machine guns

supply. Intended to give a still-air maximum range of 2221km (1380 miles), initial estimates proved somewhat optimistic, so a further additional 131.84L (29 gallon) tank was provided in the rear fuselage. For the Malta flights, only two machine guns were fitted, and fully fuelled, the aircraft exceeded their maximum take-off weight. Nonetheless, take-off was achieved without incident, and after a flight of five and a quarter hours, the first two Spitfires to fly non-stop to Malta arrived on 25 October 1942. Fifteen more Spitfires were delivered in this manner before the German retreat from El Alamein reduced the risk to the island and lessened the need to reinforce the Maltese defences.

Malta also saw a new development in the Spitfire's repertoire when locally built bomb shackles were fitted under the wings of 126 and 249 Squadron Spitfire Vcs to allow the aircraft to operate as

DESERT AIR FORCE
No. 2 Squadron of the South African Air Force (SAAF), which was part of the Desert Air Force's No. 7 Wing, became operational in Sicily on 23 August 1943, subsequently moving to new bases on the Italian mainland.

Mks V–VI: SPITFIRES OF THE WORLD

a fighter bomber with a 113.4kg (250lb) bomb under each wing. This prompted Supermarine to develop and manufacture a definitive bomb rack for 113.4kg (250lb) bombs under the wings, or an under-fuselage rack sufficient to carry a single 226.8kg (500lb) bomb.

Merlin engine variants

Merlin development had seen Rolls-Royce develop a whole family of engines with various improvements intended for different applications. The 50-series engines were built with a version of the R.A.E. Restrictor in place to solve the negative-G cut-out problem, whilst the Merlin 46 utilized an SU carburettor to inject fuel directly into the supercharger intake as an alternative solution to the problem. By the end of 1942, Spitfire V variants were utilizing a variety of different Merlins depending on their intended role. For example, versions of the Merlin 45, 50 and 55 series were fitted with cropped supercharger impellers to boost power at low altitude, and these engines were identified by an 'M' suffix. As a result, Spitfires began to receive a prefix to official mark numbers to denote their role.

This produced designations such as F Mk Va, Vb, etc, with 'F' standing for 'Fighter' as well as LF Mk Vb, Vc, etc, with 'LF' standing for 'Low Fighter' (indicating that a low-altitude Merlin 45M, 50M

Mks V–VI: SPITFIRES OF THE WORLD

SPITFIRE Mk Vc

Weight (maximum take-off): 3346kg (7420lb)
Dimensions: Length 9.12m (29ft 11in), Wingspan 11.23m (36ft 10in), Height 3.02m (9ft 10in)
Powerplant: one 1074kW (1440hp) Rolls-Royce Merlin 45 liquid cooled V-12 piston engine
Maximum speed: 597km/h (371mph)
Range: 756km (470 miles)
Ceiling: 11,460m (37,600ft)
Crew: 1
Armament: four Hispano 20mm (0.79in) cannon, or eight 7.7mm (0.303in) Browning machine guns (both options rarely fitted), or two 20mm (0.79in) Hispano cannon and four 7.7mm (0.303in) Browning machine guns in wings; up to 230kg (500lb) bomb load

SPITFIRE MK VC
Not many Spitfire Vs were still in frontline service in Western Europe by D-Day. AB509 was the personal aircraft of Wing Commander John Checketts and displays the badges of two of the three squadrons under his command, 303 (Polish) Sqn and 402 Sqn RCAF.

or 55M was fitted). For low-altitude use, the wingtips were often removed and replaced with a fairing, reducing the wingspan by 1.42m (4ft 8in) and improving rate of roll and handling at low level. This is generally referred to as the 'clipped wing' Spitfire. LF fighters with the modifications to supercharger and wingtips saw intense use during 1942 and 1943 and became somewhat careworn by the rigours of low-altitude operations, with pilots referring to their tired mounts as 'cropped, clipped and clapped'.

Possessing clipped wings did not definitively identify a Spitfire as an LF variant, however, as the modification was acceptable for medium-altitude operations as well. Many F Mk Vs with medium-altitude Merlin 45, 46, 50, 50A, 55 or 56 series engines were also fitted with clipped wings.

Mk Vc: High-altitude model

SPITFIRE MK VC
Fitted with the bulky Vokes tropical filter under the nose, this aircraft was on the strength of 54 Squadron RAF based at Darwin in the north of Australia. As a result it wears standard RAAF markings, the yellow and red elements of the RAF roundels having been painted over.

Mks V–VI: SPITFIRES OF THE WORLD

At the other end of the scale, much effort was expended on producing a high-altitude Spitfire. The first effort to bear fruit in this regard was an ad hoc modification made to a Mk Vc at Aboukir Maintenance Unit (MU), Egypt, in mid-1942. At this time, pressurized Junkers Ju 86P-2s were conducting near-daily reconnaissance flights over the Suez Canal and Alexandria at 12,192m (40,000ft) or higher, beyond the reach of any standard Allied fighter. The Spitfire selected for modification was stripped of all unnecessary equipment and armament, being fitted solely with two 1.27cm (0.50in) Browning machine guns. Locally fabricated extended wingtips were fitted along with a four-bladed propeller and a modified Merlin 46. On 24 August 1942, this aircraft was used to intercept and destroy a Junkers Ju 86P-2 at 12,802m (42,000ft), the highest successful interception of the entire war.

Mk VI

By this time, production of a high-altitude Spitfire variant, the Spitfire Mk VI, was underway in the UK, prompted by a series of nuisance raids by the Ju 86R-1 flying at sufficient altitude to render it immune to interception. The Spitfire Mk VI, later the HF Mk VI, featured a pressure cockpit contained within bulkheads fore and aft of the cabin, and utilized a Marshall blower to maintain pressure within the cockpit at 13,790 pascals (2lb per

SPITFIRE MK VI

Air was fed to the cockpit pressurisation system via an air intake beneath the engine exhausts on the starboard side. The four-bladed propeller of the Mk VI helped distinguish it from the Mk V.

sq in) higher than the outside atmosphere. The sealed canopy of the Mk VI was non-sliding but jettisonable in an emergency, though suspicions about the effectiveness of the canopy jettison system were to dog the service use of the aircraft.

A high-altitude Merlin 47 and four-bladed propellers were fitted along with extended wingtips, attached to an otherwise standard B-type wing, which increased the span to 12.24m (40ft 2in). A hundred Mk VIs were built,

Mks V–VI: SPITFIRES OF THE WORLD

SPITFIRE Mk Vb

Weight (maximum take-off): 3071kg (6525lb)

Dimensions: Length 9.12m (29ft 11in), Wingspan (clipped) 10.01m (32ft 10in), Height 3.02m (9ft 10in)

Powerplant: one 1181kW (1585hp) Rolls-Royce Merlin 50M liquid cooled V-12 piston engine

Maximum speed: 564km/h (351mph)

Range: 756km (470 miles)

Ceiling: 10,881m (35,700ft)

Crew: 1

Armament: two 20mm (0.79in) Hispano cannon and four 7.7mm (0.303in) Browning machine guns in wings; up to 230kg (500lb) bomb load

but by the time they entered service, interception of the Ju 86s that were overflying Britain by a standard Mk IX (described in the next chapter) had proved to be just possible. The Mk VI itself couldn't operate much above 10,668m (35,000ft) and was therefore unable to intercept the reconnaissance Ju 86P-2, as starkly demonstrated when six were shipped to Aboukir but were unable to match the altitude performance of the locally modified Spitfire VCs in use there.

Nonetheless, the Mk VI provided much valuable high-altitude experience to various units. Two squadrons, Nos 616 and 124, were equipped wholly with the type, which scored its first victory on 18 July 1942 when an Fw 190 was shot down. Various Fighter Command squadrons were rotated to Orkney and the Shetland Islands for rest periods, where around 12 Spitfire VIs were based, taking the aircraft on strength whilst they were there. These aircraft provided high-altitude cover against any high-altitude bombers or reconnaissance aircraft that might appear, and as a result, the Spitfire Mk VI served with many squadrons, out of all proportion to its modest production total.

Pacific theatre deployment

The latter half of 1942 saw the Spitfire deploy to the Pacific theatre for the first time with the arrival of 54 Squadron RAF equipped with Spitfire VCs at Darwin, Australia. The squadron became operational in January 1943 and scored the first 'kill' against a Japanese aircraft by a Spitfire on 6 February. Flying

SPITFIRE MK VB

The Spitfires of 249 Squadron on Malta featured this unusual colour scheme after locally sourced paint of a blue-grey hue was painted over the Middle Stone of the standard RAF tropical camouflage. AB264 was flown by Pilot Officer 'Buck' McNair, RCAF.

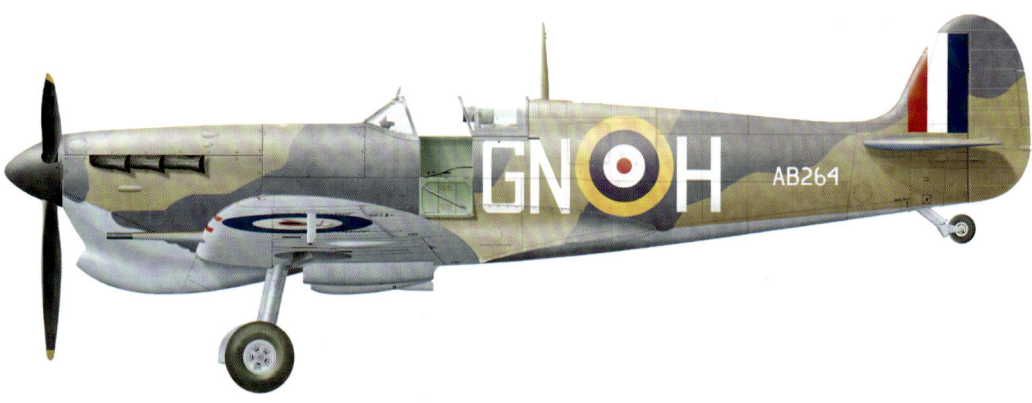

AMERICAN SQUADRON

The first major use of the Spitfire by a non-Commonwealth power took place in 1942 when the three 'Eagle' squadrons of the RAF, manned by volunteer US pilots, exchanged the British roundels of their Spitfire Vbs for American markings when they transferred to USAAF control, becoming the 334th, 335th and 336th Pursuit Squadrons of the 4th Pursuit Group. They flew their first operational mission – a sweep over the French coast – on 2 October 1942. Although these units were re-equipped with P-47s in March 1943, other USAAF squadrons were to fly Spitfire Vs in the UK and participate in the Allied landings in North Africa, 'Operation Torch', in November 1942. The 31st and 52nd Fighter Groups of the Twelfth Air Force USAAF flew Spitfire Vs alongside similarly equipped British and Commonwealth squadrons throughout the North African campaign and were a major component of the Allied fighter force that obtained absolute air supremacy over Africa by the time Axis forces surrendered in May 1943. There followed heavy engagement in the assault on Sicily and the subsequent invasion of the Italian mainland, with US squadrons exchanging their well-used Spitfire Vs for Spitfire Mk VIIIs and IXs during this period.

NEW OWNERS
These Spitfire Vbs at Shaibah Airfield, Iraq have had their RAF roundels crudely repainted into USAAF stars. The aircraft in the extreme foreground still has the RAF roundel on the starboard wing, implying the photo was taken whilst this work was ongoing.

Mks V–VI: SPITFIRES OF THE WORLD

SPITFIRE Mk Vb
Weight (maximum take-off): 3071kg (6525lb)
Dimensions: Length 9.12m (29ft 11in), Wingspan (clipped) 10.01m (32ft 10in), Height 3.02m (9ft 10in)
Powerplant: one 1181kW (1585hp) Rolls-Royce Merlin 50M liquid cooled V-12 piston engine
Maximum speed: 564km/h (351mph)
Range: 750km (470 miles)
Ceiling: 10,881m (35,700ft)
Crew: 1
Armament: two 20mm (0.79in) Hispano cannon and four 7.7mm (0.303in) Browning machine guns in wings; up to 230kg (500lb) bomb load

SOVIET SPITFIRE MK VB
Most of the Mk V Spitfires supplied to the Soviet Union saw quite intense service over Crimea but EP356, coded White 20, was retained by the Air Force Scientific Institute for flight testing.

under Australian control, 54 Squadron was joined by two Royal Australian Air Force (RAAF) Spitfire Vc squadrons to form No 1 Fighter Wing. Later the same year, further Spitfire Vcs re-equipped three RAF Hurricane squadrons on the Burma front, with the Indian Air Force also receiving Mk Vs in 1944.

Soviet transfer
A second major user of the Spitfire was to receive Spitfire Mk Vs during 1943, with the transfer of an initial batch of 143 ex-RAF Spitfire Mk Vbs to the Soviet Union. Seeing brief but intense service with the Red Army Air Force, the Spitfire was considered less than ideally suited for conditions prevailing on the Eastern Front. Its narrow track undercarriage was inadequate for dealing with the unprepared airstrips that were the

norm in the theatre, and although its handling and simple flying characteristics were appreciated by pilots, the Vb was considered to have inferior performance to the latest Yakovlev and Lavochkin fighters in use by the USSR. This was due in part to the lower-octane fuel available, which the Merlin was perfectly capable of burning without mechanical problems but which did have a deleterious effect on power output. The Spitfires were not optimized for the low-level combat that typified fighter operations over Russia, and the fact that all the Spitfires supplied were somewhat tired 'second-hand' examples didn't help either.

Two aspects in which the Spitfire Vb impressed its Soviet operators, however, were its armament – the combination of two cannon and four machine guns made it the heaviest armed single seater available to the USSR at that time – and its metal construction, which was found to be more resilient to combat damage than the largely wooden construction of domestic designs. The Spitfire Vs were replaced by

P-39 Airacobras after a few months of operations, a decision that would have bewildered non-Soviet fighter pilots who had found the P-39 inadequate for operations in the West, in stark contrast to the superlative Spitfire.

Widespread use

Other nations to fly the Spitfire Mk V operationally included South Africa, Yugoslavia, which flew the Vc on fighter-bomber duties in the Balkans until VE Day, and France, which operated Spitfire Vs both under RAF control and independently. Following the partial capitulation of Italy, 33 Spitfire Vs were made available to the Italian Co-Belligerent Air Force, these seeing service from October 1944 against their previous foes until the end of the war. Two Spitfire Vs of 20° Gruppo flew the Regia Aeronautica's last wartime mission on 5 May 1945, a reconnaissance mission over Zagreb.

Perhaps the most unlikely operator of the Spitfire V of all, however, was US Navy Cruiser Scouting Squadron Seven (VCS-7), a unit more accustomed to flying floatplanes from capital ships. VCS-7 used this profoundly land-based fighter as a spotter aircraft directing gunfire from surface vessels on D-Day and for 20 days thereafter. Drop tanks were carried on these missions, which could last up to two hours and proved extremely effective.

Experimental programs

The ready availability of the Mk V saw it utilized in some interesting experimental programmes. Spitfire Vbs were used to test dive-brakes for the Spitfire in an attempt to improve its efficacy in the dive-bombing role, but sadly none could provide sufficient deceleration, and this development was abandoned. More successful was the work by Professor A R Hill with a Spitfire Vb modified by Westland to perform the first tentative experiments in boundary layer control, which would pay great dividends for much later aircraft designs. A rather more dramatic scheme resulted in the adaptation of the Spitfire V as a glider tug, the idea being that a squadron of fighters could each tow a Hotspur glider carrying ground crew, spares and fuel to allow the Spitfires to begin operations from an advanced airstrip. The feasibility of the concept was proved but it was never used operationally.

The Spitfire V's excellent flying characteristics also saw it used as a development airframe for a most unexpected operator: the Luftwaffe. A Spitfire Vb of 131 squadron that had force-landed in France in November 1942 was re-engined by Messerschmitt with a Daimler-Benz DB 605A engine for work on the cooling system and used in comparative trials against the Bf 109G. Later it was fitted with a DB 601A engine, but was ultimately destroyed in an Allied bombing raid.

PORTUGUESE SPITFIRE MK VB
Neutral Portugal received 33 Spitfire Mk Vs from RAF stocks during 1943 as part payment for Allied use of airfield facilities for anti-submarine aircraft in the Azores. The Spitfires were bolstered by further Mk Vs supplied after the end of the conflict.

Mks VII–XVI: Ultimate Merlin Marks

'Whatever these strange fighters were, they gave us a hard time of it. They were faster in a zoom climb than the Spitfire, far more stable in a vertical dive, and they turned better than the Messerschmitt.' Throughout 1941, the Spitfire Mk V was arguably the finest fighter in the world, its only real rival being the Messerschmitt 109F (which possessed a similar performance and armament). However, late in 1941, a new German fighter appeared that was to have a profound effect on the development of the Spitfire: the Focke-Wulf Fw 190.

D-DAY INVASION STRIPES
MK959, a Spitfire LF Mk IXC is a genuine combat veteran having served with 302 (Polish) Squadron, 165 Squadron and the French-manned 329 Squadron, whose colours it wears here, and with whom it flew sorties over the Normandy beachhead in June 1944.

Mks VII–XVI: Ultimate Merlin Marks

SPITFIRE F MK VII

Wearing unusually small non-standard 'invasion stripes', this Mk VII operating with 154 Squadron in August 1944 is finished in standard RAF high-altitude camouflage of medium sea grey above and PRU blue undersides.

SPITFIRE Mk VII

Weight (maximum take-off): 3583kg (7900lb)
Dimensions: Length 9.54m (31ft 4in), Wingspan 11.23m (36ft 10in) or 12.24m (40ft 2in) with extended wingtips, Height 3.86m (12ft 8in)
Powerplant: one 1167kW (1560hp) Rolls-Royce Merlin 61 liquid cooled V-12 piston engine
Maximum speed: 657km/h (408mph)
Range: 1062km (660 miles)
Ceiling: 13,106m (43,000ft)
Crew: 1
Armament: two 20mm (0.79in) Hispano cannon and four 7.7mm (0.303in) Browning machine guns in wings

The description above, of encountering the Fw 190 for the first time, was written by James 'Johnnie' Johnson, the RAF's most successful Spitfire pilot in air combat, and his views were echoed in a letter written by Hugh Dowding, the former head of Fighter Command, to Winston Churchill in 1942: 'It is not only in performance that the Fw 190 has the Spitfire beaten. It has superior hitting power as well.' Dowding adds that 'the Spitfire has been moribund for two years and died when the Fw 190 made its appearance'. While it was true that the Fw 190 was superior in every measurable parameter to the Spitfire V except turning circle, Dowding's assessment that the Spitfire programme was as good as finished would prove massively premature.

Developing the Mk VII & Mk VIII

When the Fw 190 appeared, Joe Smith's team at Supermarine had in fact been working for some time on a considerably superior fighter to the Mk V, or rather two: the Mk VII and Mk VIII. Developed in parallel, they shared a swathe of improvements but differed in that the Mk VII was intended for the high-altitude role and featured a pressurized cockpit. Neither aircraft was expected to be ready for several months when the Fw 190 appeared, however, and another solution was desperately needed. Luckily for Fighter Command, Rolls-Royce had been working on a new engine – the Merlin 60 series – intended for use in the high-altitude Vickers Wellington VI bomber. This aircraft featured a two-stage, two-speed supercharger design – a result of the inspired work of engineer Stanley Hooker – that

Opposite: SPITFIRE F MK IXC Ordered in October 1940 as a Spitfire Mk I, BS459 was one of the earliest Mk IXs to leave Supermarine's Eastleigh factory. Completed in September 1942, it was issued to No. 306 'Torunski' Squadron, a Polish unit based at RAF Northolt. Engaged in daylight sweeps over Europe, the hapless aircraft and its pilot failed to return from operations on 26 January 1943 and are believed to have collided with sister aircraft BS241 over the English Channel.

Mks VII–XVI: ULTIMATE MERLIN MARKS

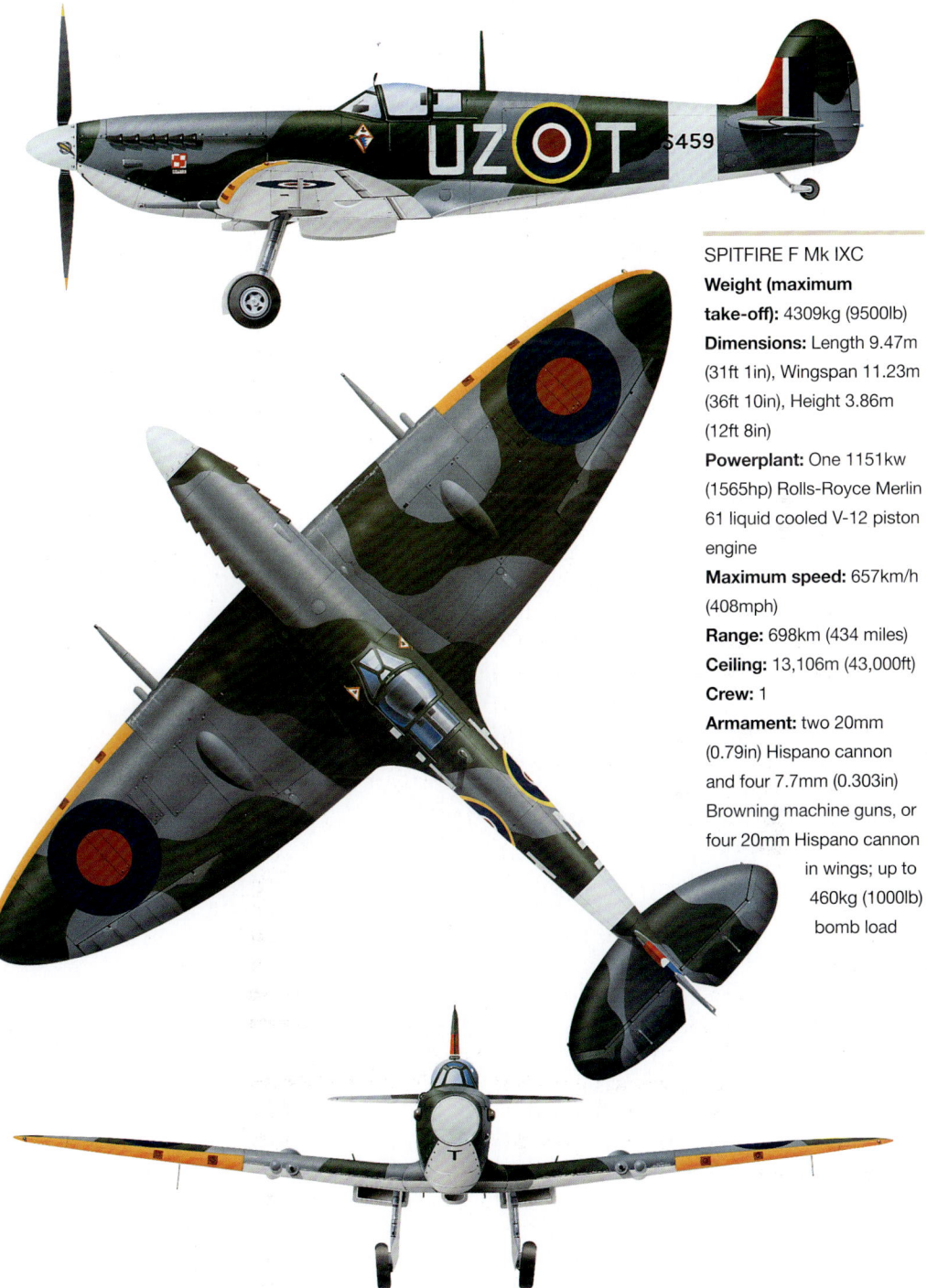

SPITFIRE F Mk IXC
Weight (maximum take-off): 4309kg (9500lb)
Dimensions: Length 9.47m (31ft 1in), Wingspan 11.23m (36ft 10in), Height 3.86m (12ft 8in)
Powerplant: One 1151kw (1565hp) Rolls-Royce Merlin 61 liquid cooled V-12 piston engine
Maximum speed: 657km/h (408mph)
Range: 698km (434 miles)
Ceiling: 13,106m (43,000ft)
Crew: 1
Armament: two 20mm (0.79in) Hispano cannon and four 7.7mm (0.303in) Browning machine guns, or four 20mm Hispano cannon in wings; up to 460kg (1000lb) bomb load

65

Mks VII–XVI: Ultimate Merlin Marks

SPITFIRE F MK IXC
MH819 was on the strength of Czechoslovak 310 Squadron and is depicted as it appeared in mid June 1944 when based at Appledram Airbase, West Sussex. As the serial number was obscured by the D-Day stripes, the ground crew repainted it on the tail fin.

SPITFIRE F MK IXC
Fitted with the Aboukir filter and unusually finished in high-altitude colours, this Mk IX flew with 451 Squadron RCAF, under RAF control. Initially based at Casablanca, the squadron had moved to Poretta on Corsica by May 1944.

SPITFIRE F MK IXC
BS152 was flown by Flying Officer Lorne Cameron of 402 Squadron RCAF, based at RAF Kenley in Surrey. On 27 February 1943, his 21st birthday, Cameron used this aircraft to shoot down a Fw 190A-4 of JG 26.

SPITFIRE F Mk IXC
Weight (maximum take-off): 4309kg (9500lb)
Dimensions: Length 9.47m (31ft 1in), Wingspan 11.23m (36ft 10in), Height 3.86m (12ft 8in)
Powerplant: One 1151kw (1565hp) Rolls-Royce Merlin 61 liquid cooled V-12 piston engine
Maximum speed: 657km/h (408mph)
Range: 698km (434 miles)
Ceiling: 13,106m (43,000ft)
Crew: 1
Armament: two 20mm (0.79in) Hispano cannon and four 7.7mm (0.303in) Browning machine guns, up to 460kg (1000lb) bomb load

Mks VII–XVI: ULTIMATE MERLIN MARKS

was able to maintain a boost level of 9lb per sq in (62,053 pascals) up to 9144m (30,000ft). As a result, power output at that altitude was effectively doubled from 500hp to 1000hp.

Ernest Hives, the chairman of Rolls-Royce, appears to have been the first person to have realized that this engine might be suitable for the Spitfire, and notwithstanding the increased length and additional cooling required for the new Merlin, Rolls-Royce sought and obtained Air Ministry approval to attempt to fit a Merlin 60 in a Spitfire airframe. The actual airframe selected for conversion was the prototype Spitfire Mk III, which was duly modified to receive the Merlin 60, a Rotol four-bladed propeller and twin underwing radiators, entering flight testing in September 1941. A Spitfire Mk IA was converted in a similar fashion and fitted with a Merlin 61, the first production variant of the 60 series. Initial testing proved encouraging, and two Spitfire Vcs were fitted with Merlin 61s to become the prototypes of the Spitfire Mk IX, though these were initially known simply as Mk Vc (Merlin 61).

Mk IX: Mass-production model

It was the appearance of the Fw 190 and its obvious superiority over the Spitfire V that prompted the decision to mass-produce the Mk IX, which was essentially a Mk V airframe with a new engine shoehorned into it, rather than wait for the more sophisticated Mk VII and Mk VIII. Once again, much as with the Mk III and Mk V, an extemporized conversion had sidelined a more advanced design.

Unlike the Mk III, however, both the VII and VIII would eventually enter production, though in lesser numbers than the mass-produced Mk IX. Against the Mk V, the IX was 112.7km/h (70mph) faster and its combat altitude was increased by 3048m (10,000ft).

The Air Fighting Development Unit (AFDU) tested a captured Fw 190A-3 against the Spitfire Mk IX in July 1942 and found the Fw 190 to be slower at medium and high altitudes, although its rate of climb was superior.

SPITFIRE F MK IXC
The personal aircraft of the most successful Spitfire pilot of all, Wing Commander James Edgar 'Johnnie' Johnson. Built as a Mk IXC, MK392 was later fitted with the 'E' wing and was used by Johnson to score 12 of his total of 38 victories.

It was ascertained that the Mk IX did not enjoy the same advantage over the Fw 190 as the German aircraft did over the Mk V, but it did at least restore parity or a slight superiority over the Focke-Wulf machine. Production was ordered to begin at the CBAF as soon as the supply of Merlin 61 engines could be guaranteed. In the meantime, Rolls-Royce was contracted to convert 282 Mk Vc airframes to Mk IX standard.

This work was undertaken with such urgency that the first Mk IXs were delivered to 64 Squadron at Hornchurch during June 1942, and the Spitfire Mk IX entered combat for the first time (and in fine style) on 30 July, intercepting a formation of 15 Fw 190s off the French coast and shooting down three of them. Following the advent of the

Mks VII–XVI: ULTIMATE MERLIN MARKS

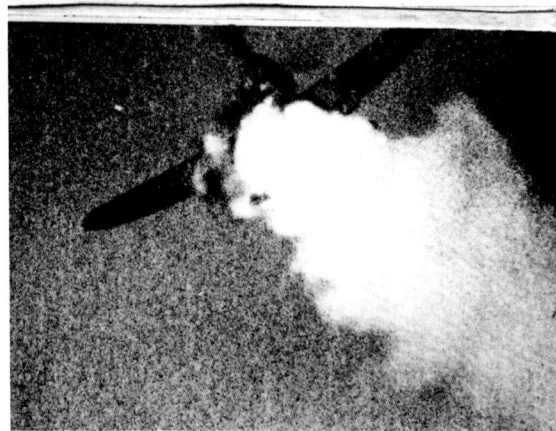

AERIAL KILL
A Bf 109G succumbs to the guns of RCAF Sergeant G Shouldice's Spitfire Mk IXB. These gun camera stills, of unusual clarity for the era, show the devastating effect of the Hispano 20mm (0.79in) cannon.

Spitfire Mk IX, although various aircraft would at times achieve broadly similar performance to this and later Spitfires, no German piston engine fighter would again threaten to outclass it, and the Spitfire maintained an edge over later models of both the Bf 109 and Fw 190 until the war's end. UK-based squadrons were given priority for Mk IX deliveries and units were re-equipped from Spitfire Vs as production built up over the course of 1942, whilst it was decided that squadrons in the Middle and Far East would wait to receive the Mk VIII when it became available – though Mk IXs would in due course see service in Italy alongside the Mk VIII.

'E' Wing

The Mk IX initially featured the C-type wing of its Mk Vc forebear, but this was replaced later in production with the 'E' wing, which dispensed with the two outboard 7.7mm (0.303in) machine guns and instead featured one 20mm (0.79in) cannon outboard and one 1.27cm (0.50in) Browning machine-gun immediately inboard on each side. As the pattern of air warfare changed over Western Europe

Mks VII–XVI: ULTIMATE MERLIN MARKS

and the Mk IX found itself used less for its initial fighter sweep and bomber escort missions, wing bomb racks allowing for the carriage of two 113.4kg (250lb) bombs under the wings or one 226.8kg (500lb) bomb under the fuselage were provided, much as had been the case with the Mk V. Likewise, the fighting altitude of the Mk IX was dictated by what version of Merlin engine was fitted and designation prefixes were introduced: LF Mk IX for aircraft fitted with the Merlin 65; F Mk IX for aircraft with the Merlin 61 or 63; and HF Mk IX for machines powered by the Merlin 70. Although the standard span wing was utilized on most Mk IXs, the increasing importance of low-level missions saw many aircraft receive clipped wings.

As well as possessing a nose some 17.8cm (7in) longer than the Spitfire V, later production Mk IXs also introduced a broad chord, pointed tip rudder that changed the profile outline of the aircraft, and many earlier airframes were fitted with the new rudder retrospectively. A more dramatic outline change came with a switch on the CBAF production line to a 'low back' fuselage which had been trialled on a Mk VIII airframe and allowed for the fitting of a 360-degree view teardrop canopy. This did nothing for the aesthetics of the aircraft but proved immensely beneficial to the pilot. Another change, barely noticeable externally, saw a rear fuselage tank of 327.32L (72 gallons) added within the rear fuselage, usefully extending the range. Aircraft so-equipped could be identified by the addition of a filler cap in the port side of the Perspex panel immediately behind the cockpit canopy.

Reorganization

In the lead-up to D-Day, the organization of RAF Fighter Command's squadrons in the UK was changed. Home defence became the mission of the Air Defence of Great Britain (ADGB), and the squadrons tasked with supporting the upcoming invasion of the European mainland were assigned to the 2nd Tactical Air Force (2TAF). At the end of 1943, ADGB possessed 22 Spitfire squadrons, and 2TAF, 34. Unsurprisingly, given the importance of the upcoming Operation Overlord, 2TAF had priority with regard to re-equipping Spitfire squadrons. As a result, some ADGB squadrons were still operating the Spitfire V until mid-1944.

Entering service

By this time, the Spitfire Mk VII and Mk VIII had both entered service. As the more specialized of the pair, the Mk VII was built only in small numbers, with a mere 140 coming off the assembly line at Supermarine. Befitting its high-altitude role, the Mk VII featured the extended wingtips first seen on the Mk VI, but there were also structural changes in both wing and fuselage. The capacity of the fuel tanks in the forward fuselage was increased, and tanks were added to the leading edge of each wing, adding another 127.29L (28 gallons). The Mk VII featured the broad chord rudder with the pointed tip (as fitted to some Mk IXs), along with the twin underwing radiator arrangement that had become necessary to cool the Merlin 60 series and the supercharger intercooler. All Mk VIIs were fitted with a low-drag Aero-Vee dust filter from the factory, a fitment that became standard on later production Spitfire IXs also. Additionally, for the first time on a production Spitfire, a retractable tailwheel was fitted. To enable cabin pressurization, a Marshall Mk XII supercharger was fitted on the starboard side of the Merlin 61 crankcase to maintain cockpit pressure, and the Spitfire Mk VII benefitted from a much-improved canopy sealing system when compared to the Mk VI, with the hood of later production examples capable of sliding open or closed as on a conventional machine. Most Mk VIIs featured the Merlin 64 engine, though a few utilized the special high-altitude Merlin 71.

First 'kill'

When role prefixes were introduced, the Merlin 71 examples became the HF Mk VII, while all others became F Mk VIIs. Deliveries began in September 1942, with the first three aircraft going to Fighter Command's High Altitude Flight

69

BEER RUN

As 2TAF followed the Allied armies across Europe in the weeks following D-Day, the Spitfire IX performed perhaps the most bizarre mission of the entire Spitfire programme: operating as a beer transport. In response to a shortage of beer in France as the Allies advanced, several British breweries offered to supply free beer for the troops overseas, the only difficulty being how to get it there. Initially, cleaned external fuel tanks were used (jokingly referred to as 'Modification XXX'), but this imparted a slight metallic taste when it arrived – yet despite this, the beer flights increased in regularity, and some Hawker Typhoons were utilized in this role as well.

Eventually, at least one Spitfire Mk IX was equipped to carry wooden beer barrels on its underwing bomb wracks, which meant that the beer tasted fine and was also pleasantly chilled by the flight over the Channel.

At best, semi-official, the beer flights were eventually brought to a halt by HM Customs and Excise, who warned that the breweries were in breach of the law as providing free beer to the troops in France bypassed export tax, and with that, the participating Spitfire Mk IXs returned to their more conventional combat roles.

BARREL LOAD
A Mk IX Spitfire carries its precious alcoholic cargo to the troops in France. One wonders what would have happened had the aircraft been intercepted: it is not recorded whether the barrels could be jettisoned.

at Northolt that had been set up to intercept the ongoing Ju 86P incursions over British airspace. However, no interceptions were made, and the flight was absorbed into 124 Squadron, who were to score the Mk VII's first 'kill' on 15 May 1943 when an Fw 190 was destroyed at 11,582m (38,000ft) over Plymouth. Two further squadrons, Nos 616 and 131, were equipped with the type, although the lack of high-altitude targets saw them being used in support of the D-Day landings, with at least one squadron discarding the extended wingtips since all their operations were at medium or

Mks VII–XVI: ULTIMATE MERLIN MARKS

SPITFIRE HF MK VIII
This Mk VIII was on the strength of 457 Squadron RAAF, known as the 'Grey Nurse' squadron as its aircraft were decorated with the mouth of the Grey Nurse shark. The Mk VIIIs were received when the squadron was based at Livingstone in the Northern Territory. In January 1945 it relocated to Morotai in the Dutch East Indies (now Indonesia).

SPITFIRE Mk VIII
Weight (maximum take-off): 3638kg (8020lb)
Dimensions: Length 9.54m (31ft 4in), Wingspan 11.23m (36ft 10in), Height 3.86m (12ft 8in)
Powerplant: One 1275kW (1710hp) Rolls-Royce Merlin 63 liquid cooled V-12 piston engine
Maximum speed: 657km/h (408mph)
Range: 1062km (660 miles)
Ceiling: 13,106m (43000ft)
Crew: 1
Armament: Two 20mm (0.79in) Hispano cannon and four 7.7mm (0.303in) Browning machine guns; up to 460kg (1000lb) bomb load

Mks VII–XVI: ULTIMATE MERLIN MARKS

SPITFIRE Mk VIII
Weight (maximum take-off): 3638kg (8020lb)
Dimensions: Length 9.54m (31ft 4in), Wingspan 11.23m (36ft 10in), Height 3.86m (12ft 8in)
Powerplant: One 1275kW (1710hp) Rolls-Royce Merlin 63 liquid cooled V-12 piston engine
Maximum speed: 657km/h (408mph)
Range: 1062km (660 miles)
Ceiling: 13,106m (43000ft)
Crew: 1
Armament: Two 20mm (0.79in) Hispano cannon and four 7.7mm (0.303in) Browning machine guns

low altitude. The Mk VII was also used by resting Spitfire squadrons in the Orkneys, just as had been the case with the Mk VI. Only 131 Squadron continued using the Mk VII operationally after August 1944, passing on their aircraft to 154 Squadron, which flew them until the end of hostilities.

Mk VIII in action
Much more widespread use was made of the Mk VII's lower-level counterpart: the Mk VIII. With all the same modifications and improvements as the Mk VII but lacking the cockpit pressurization equipment, the Mk VIII was the definitive Merlin-powered Spitfire variant. Supermarine chief test pilot Jeffrey Quill stated that it was the best Spitfire of all from a flying point of view when fitted with normal wingtips, but that the extended wingtips ruined the handling and offered no great advantage to altitude capability. Nonetheless, some early Mk VIIIs were fitted with the high-altitude

SPITFIRE HF MK VIII
Operating as part of 80 Fighter wing of the RAAF's 1st Tactical Air Force, this Mk VIII of 457 Squadron was flown by Flight Lieutenant Edward Sly, ending the war at Labuan in Borneo. In 2007, on his 90th birthday, Sly subsequently flew solo again for the first time since 1945.

SPITFIRE F Mk VIII
'Betty Jane' was the personal aircraft of Colonel Charles M McCorkle, CO of the 31st Fighter Group based at Castel Volturno, Italy in March 1944. McCorkle scored the first five of his 11 victories with the Spitfire VIII and IX, the remainder flying a P-51 Mustang, also named 'Betty Jane'.

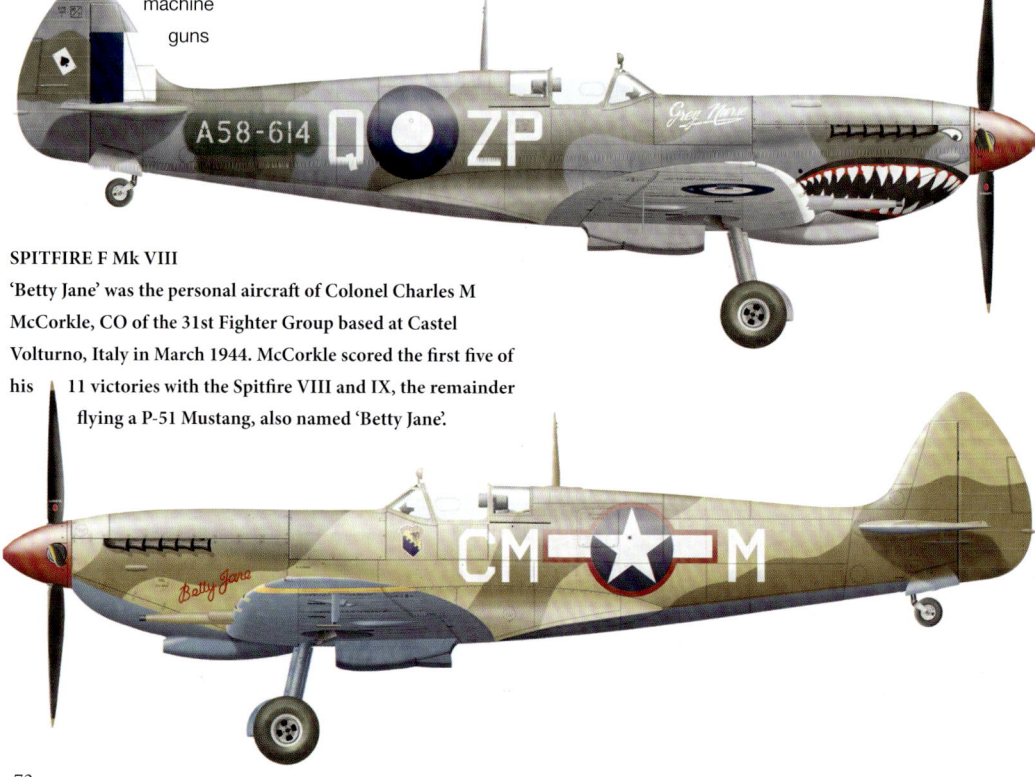

Mks VII–XVI: ULTIMATE MERLIN MARKS

COMMANDING OFFICER Air Vice Marshal Harry Broadhurst, AOC Desert Air Force, taxis his personally-coded Mk VIII past the remains of airship hangers at Taranto, Italy in the summer of 1943.

wingtips. Like the Mk IX, the Mk VIII received subtype prefixes depending on the type of Merlin engine fitted. Those with the low altitude rated Merlin 66 became LF Mk VIIIs, the standard Mk 61 or 63 Merlin aircraft became F Mk VIIIs and the Spitfire HF Mk VIII was fitted with the Merlin 70. Wingtip configuration varied between the different subtypes, the allotted designation always being derived from the engine fitted.

Italian campaign

The first Mk VIIIs were issued to Nos 92 and 145 Squadrons in the Middle East in mid-1943 – just as the North African campaign was ending. Subsequently, these and other similarly equipped units would see much action in the invasion of Sicily and the advance up the toe of Italy. Both US and Free French squadrons flew the Spitfire VIII and Spitfire Mk IX in Italy. The French units would later take their aircraft to the south of France in support of Operation Dragoon. The American 31st and 52nd Fighter Groups would eventually trade in their Spitfires for P-51B Mustangs in March 1944, though not to the universal delight of their pilots: a mock dogfight was staged between a Mustang and a Spitfire Mk IX, and the pilot of the new P-51B found himself completely outmanoeuvred. In total, the 600 Spitfires made available to the USAAF under reverse Lend-Lease arrangements accounted for 379 enemy aircraft, and several American pilots achieved 'ace' status in the Supermarine fighter. The Italian campaign also saw a greater emphasis on ground attack missions and the Mk VIII was equipped with a new type of bomb rack carrying two 226.8kg (500lb) bombs side by side under the centre section. An automatic delay of one-fifth of a second between the bombs being released prevented either weapon from fouling the other.

Far East service

Further east, Spitfire Mk VIIIs began to supersede Spitfire Vs in Burma, with some 10 RAF squadrons re-equipping on the Mk VIII and flying operationally

73

MISSION ACCOMPLISHED
Following a successful fighter sweep against the Japanese, pilots Flight Lieutenant DA Pidgeon, Warrant Officer WG Yates and Flight Sergeant RH Whittle of No. 607 Squadron cross the rainsoaked airfield at Imphal, India, for a post-mission debrief – their Spitfire Mk VIIIs can be seen in the background.

Mks VII–XVI: Ultimate Merlin Marks

SPITFIRE Mk VIII

Weight (maximum take-off): 3638kg (8020lb)
Dimensions: Length 9.54m (31ft 4in), Wingspan 11.23m (36ft 10in), Height 3.86m (12ft 8in)
Powerplant: One 1275kW (1710hp) Rolls-Royce Merlin 63 liquid cooled V-12 piston engine
Maximum speed: 657km/h (408mph)
Range: 1062km (660 miles)
Ceiling: 13,106m (43000ft)
Crew: 1
Armament: Two 20mm (0.79in) Hispano cannon and four 7.7mm (0.303in) Browning machine guns; up to 460kg (1000lb) bomb load

SPITFIRE F MK VIII

Displaying 152 Squadron's striking leaping panther emblem and the South East Asia Command blue roundels, this Mk VIII was based at Sinthe in Burma. Flying Officer Len Smith shot down a Nakajima Ki-43 in this aircraft at the end of 1944, adding to the four kills he had already attained over Tunisia and Italy.

before VJ Day. Another major user of the Mk VIII was Australia. No 1 Fighter Wing, consisting of Nos 452 and 457 Squadrons RAAF and 54 Squadron RAF, re-equipped with Spitfire Mk VIIIs in 1944. Further squadrons would follow, with the Spitfire Mk VIII deploying for operations in New Guinea. The Spitfire would eventually become the top-scoring Australian fighter in terms of air-to-air victories against the Japanese, accounting for 71 confirmed and 17 probable kills. Eventually, the Mk VIII would become the third most produced Spitfire type, after the Mk V and the Mk IX, with 1658 produced – all by Supermarine – while the CBAF continued with the production of Mk IXs.

Mk XVI

The final Merlin-powered Spitfire variant would appear in late 1944 as the Spitfire Mk XVI. This version was essentially identical to the Mk IX in all respects except that it was fitted with a US-built Packard Merlin 266, equivalent to the British Merlin 66. The Merlins constructed by Packard were built using American standard measurements throughout and thus not all spare parts were interchangeable between the engines built by Rolls-Royce and Packard. The engine also differed somewhat in its supercharger and cooling systems. The new Mark number was introduced to make it clear that this aircraft required a different parts supply chain to the Mk IXs it resembled and supplemented. All Mk XVIs were built at the CBAF, and as the Merlin 266 was a low-altitude engine, all were LF Mk XVIs. Despite production beginning so late in the war, 1054 were produced by the CBAF, which

Mks VII–XVI: Ultimate Merlin Marks

ALLIES IN ITALY
The two most famous Merlin-powered fighters share the ramp at Nettuno airfield, 60km (40 miles) south of Rome. A Spitfire F Mk IXC of 43 Squadron ('The Fighting Cocks') takes on fuel next to a parked P-51C Mustang of the US 15th Air Force.

had long since overcome its teething troubles and was able to deliver new aircraft at a prodigious rate.

Initially appearing with the 'C' wing and a standard canopy and rudder, by 1945, production Mk XVIs featured the 'E' wing along with broad chord rudder and cut-down rear fuselage. The majority of Mk XVIs that saw service were also equipped with clipped wingtips. Mk XVIs began operations from November 1944, with five RAF squadrons, and were used primarily to attack V-2 sites in the Netherlands. In order to attack such small targets, the Mk XVI Spitfires dived from around 2438m (8000ft) in a 70-degree dive, releasing their bombs at 914m (3000ft).

Several late Merlin variants were used in various wartime trials, including several that were engaged in experiments intended to address the ongoing problem of the Spitfire's inadequate range. Two US-style teardrop-shaped tanks were fitted under the wings containing 281.86L (62 gallons) each. Sadly, separation problems when the tanks were jettisoned were never overcome, and the tanks were not adopted for operational service. Another experimental external fuel tank consisted of a long, streamlined tank of 909.22L (200 gallons) that fitted underneath the fuselage, but a smaller, streamlined tank of 227.3L (50 gallons) capacity eventually proved successful and began to replace the flush-fitting fuel tanks previously developed in 1943.

In the summer of 1944, two Spitfire Mk IXs were modified at Wright Field in the USA with internal tankage for 909.22L (200 gallons) of fuel.

Equipped with a P-51-style drop tank under each wing, this allowed them to make a non-stop transatlantic flight from Newfoundland to Northern Ireland. By this time, enough P-51 Mustangs were available to make further long-range development of the Spitfire purely an academic exercise, and no such impressive range was ever achievable by an operational Spitfire fighter (though photo reconnaissance variants were another matter entirely.

Rocket armament
Various armament experiments were also undertaken, mostly intended to improve the aircraft as a ground-attack asset. Several trial installations of rocket projectiles

Mks VII–XVI: ULTIMATE MERLIN MARKS

SPITFIRE LF MK XVI

Finished in the colours of Group Captain Aleksander Gabszewicz, CO of No. 131 (Polish) Wing, and featuring his personal 'boxing dog' motif on the engine cowling, TE311 is actually a preserved aircraft operated by the Battle of Britain Memorial Flight, which flew again after a 10-year restoration in 2002.

SPITFIRE LF Mk XVI

Weight (maximum take-off): 3900kg (8598lb)
Dimensions: Length 9.54m (31ft 4in), Wingspan (clipped) 9.93m (32ft 10in), Height 3.86m (12ft 8in)
Powerplant: One 1287kW (1750hp) Packard Merlin 226 liquid cooled V-12 piston engine
Maximum speed: 660km/h (405mph)
Range: 698km (434 miles)
Ceiling: 12,954m (42,500ft)
Crew: 1
Armament: two 20mm (0.79in) Hispano cannon and two 12.7mm (0.5in) Browning machine guns; up to 460kg (1000lb) bomb load

SPITFIRE LF MK XVI

Delivered in August 1945, SL721 was used as the personal aircraft of Air Vice Marshal James M Robb between 1946 and 1948, who had it painted in this distinctive overall blue colour scheme.

were made on Spitfire IXs and XVIs, including the American M10 launching tube for 11.4cm (4.5in) rockets. These were arranged in a triple mounting under each wing, as well as four standard 7.6cm (3in) rocket launching rails under each wing, as used to great effect by the Hawker Typhoon. Particularly impressive was the experimental 'Triplex' rocket consisting of a single 18.3cm (7.2in) howitzer shell propelled towards its target by three 7.6cm (3in) rocket motors. However, only one unit – 74 Squadron – is believed to have flown rocket-firing Spitfires on operations before the end of hostilities, utilizing both Mk IXs and XVIs fitted with a single 7.6cm (3in) rocket under each wing.

Opposite: FLOATPLANE CONVERSION

The single Mk IX-based Spitfire floatplane gets 'on the step' during flight trials in the UK. As well as being visually distinctive, the Spitfire floatplane was noted for producing a noise like a church organ in flight.

78

SPITFIRE FLOATPLANE

The possibility of operating a Spitfire floatplane for operations where airfields were unavailable was entertained for a considerable length of time. The German invasion of Norway in April 1940 was the initial spur for considerable RAF interest in the possibility of developing a floatplane fighter. A Mk I Spitfire was duly used for a trial installation of Roc floats but was never flown in this form. The end of the Norwegian campaign saw interest in the project wane, and the urgent requirement for standard Spitfires saw the aircraft returned to landplane configuration.

The fortunes of the floatplane Spitfire revived, however, in 1942, following the entry of Japan into the war. Japan was notably the only nation to use floatplane fighters operationally, and the ability to operate combat aircraft far from land bases was regarded as clearly advantageous for the vast oceanic combat arenas of the Pacific. Folland Aircraft Ltd was assigned the task of constructing a pair of Supermarine-designed floats, each 7.8m (25ft 7in) long, which were fitted to a Spitfire Mk Vb that was modified with a four-bladed propeller, a prominent ventral fin, and a carburettor intake extended forwards under the nose to prevent water ingestion. In this form, the floatplane was found to possess inadequate directional stability, and the standard tailfin was increased in area to correct this. Despite the addition of the large floats and an increase in weight of 430.9kg (950lb), the floatplane was found to be only around 72.4km/h (45mph) slower than the landplane Mk Vb, and Folland was contracted to build 12 sets of floats and modify two more Spitfire Vbs. In October 1943, these two floatplanes, along with the prototype, were shipped to Egypt, where they were flown from the Great Bitter Lake in preparation for planned operations to intercept German transport aircraft in the Dodecanese Islands. Unfortunately, the German capture of the British-held islands of Kos and Leros saw this plan abandoned.

One final Spitfire was converted to a floatplane in the spring of 1944 for possible operations in the Pacific. Derived from the Mk IX, it utilized the same conversion parts as the earlier Mk V conversions. In this form, the Spitfire floatplane was faster than a standard Hurricane and possessed good handling, both in the air and on the water – a testament to Supermarine's long experience with waterborne aircraft. Official interest in the concept waned, however, the floatplane programme was dropped, and the sole Mk IX conversion was returned to landplane form, drawing an end to this intriguing Spitfire development.

Mks XII–XVIII: Griffon Engine Variants

In July 1942, at the Royal Aircraft Establishment, Farnborough, a demonstration of the RAF's new Hawker Typhoon was organized for various officials from the Air Ministry. The culmination of the demonstration was to be a low-level race between the Typhoon and a captured example of the Focke-Wulf Fw 190, widely believed at the time to be the fastest fighter in the world. To add some context to the proceedings, Supermarine were asked to supply a Spitfire as the third competitor in the race, its role simply to come last and demonstrate how much faster the other two were.

TEST FLIGHT
The first of the Griffon-engined Spitfires to enter service, the Mk XII was the only variant to have the single-stage engine, asymmetric radiators and four-bladed propeller. This example is seen on a test flight prior to squadron delivery.

On the day of the race, however, as the aircraft streaked past the finishing line at full throttle in front of the assembled VIPs, the finishing order was the opposite of that which had been expected: the Fw 190 was last, the Typhoon second, and the Spitfire was accelerating away from both of them. As Jeffrey Quill, the pilot of the Spitfire in question, put it in his memoirs, this result 'certainly put the cat amongst the pigeons'.

Not all was as it seemed, however. Quill had not flown a typical production Spitfire to Farnborough, but a prototype fitted with the potent new Rolls-Royce Griffon engine, with the express purpose of showing what it could do. As he stated later, 'Nobody said what sort of Spitfire I should bring. Just a Spitfire.' Within eight days of the race, the Air Ministry had placed an order for the first of 100 Mk XIIs powered by the Griffon.

Mk IV Griffon engine

Work on a Spitfire with a Griffon engine – the Mk IV – began at Supermarine in late 1939 but progressed slowly due to difficulties in modifying and strengthening the airframe for the increased weight and power of the new engine. Wartime pressure on the development and production of the early Merlin variants also saw the Griffon Spitfire programme sidelined, development work ceasing for a time during early 1940 (on the orders of Lord Beaverbrook at the Ministry of Aircraft Production) to allow maximum effort on the Merlin. The Mk IV eventually made its first flight in late 1941. Apart from the obvious changes to the nose contours of the Spitfire allowing for the larger engine to be fitted, the Mk IV/XX featured an engine thrust line lowered by two degrees, improving view over the nose, which had increased in length by 25cm (9.84in) over the Mk IX, but the rest of the aircraft was largely unaltered externally.

By the time the Mk IV entered flight test, a photo-reconnaissance Merlin-powered PR Mk IV had been developed, and to avoid confusion, the Griffon-powered Mk IV was redesignated the Mk XX. Meanwhile, significant improvements to the Merlin, culminating in the two-stage, two-speed supercharger-equipped 61 Series as fitted to the Spitfire Mk IX, rendered the development of the Griffon seemingly less urgent. It was the low-altitude performance of the Fw 190 that would change the fortunes of the Griffon Spitfire. Whilst the Mk IX, developed specifically to counter the Focke-Wulf fighter, was well able to deliver a performance as good as or superior to the Fw 190 at medium and high altitude, that advantage fell off at low levels, and from the autumn of 1942, the

GRIFFON V-12 ENGINE

The Griffon, like the Merlin, was a liquid-cooled V-12 engine, which had been in development by Rolls-Royce since 1938. Its origin lay in a request from the Fleet Air Arm for a larger version of the Merlin optimized for naval use. The engine was required to deliver high power at low altitude, possess excellent reliability, and be easy to maintain and service. Despite its greater capacity of 37 litres (8.14 gallons), making it a third larger by volume than the Merlin, careful design meant that the frontal area was only 0.73 sq m (7.9 sq ft) as opposed to the Merlin's 0.7 sq m (7.5 sq ft). The prospect of an engine delivering nearly double the power of the Merlin but with a minimal size penalty was obviously attractive, and despite its Naval origin, officials at the Air Ministry were enquiring about the possibility of fitting the Griffon to the Spitfire even before the engine was first run on 30 November 1939. One seemingly unimportant quirk of the new engine was that it rotated the propeller in the opposite direction to the Merlin, a design choice made to bring Rolls-Royce in line with other British engine manufacturers and simplify the supply of airscrews. This alteration would prove to have serious consequences when the aircraft entered service, particularly when it was eventually taken to sea.

Mks XII–XVIII: GRIFFON ENGINE VARIANTS

Luftwaffe had initiated low-level fighter-bomber raids, flying below radar coverage, against targets on the south coast of England that were proving maddeningly difficult to intercept.

Curiously, the 'tip-and-run' Fw 190 raids also proved decisive for the fortunes of the Hawker Typhoon. When the raids started, the Typhoon programme was suffering from serious development difficulties and was in danger of cancellation. However, as the Typhoon was the only RAF fighter in service with the performance necessary to have a chance of intercepting the Focke-Wulf at low altitude, the troubled Hawker fighter suddenly found itself in urgent demand to fly standing patrols over the channel coast. The somewhat neglected Griffon-powered Spitfire also offered excellent low-altitude performance, as demonstrated in the race at Farnborough.

Mk XII

The first Griffon Spitfire to enter production was the Mk XII, all examples of which were modified Mk Vc and Mk VIII airframes. As a result, some Mk XIIs would feature a retractable tailwheel and others would not, depending on which type of airframe was the basis for that specific conversion. As with the Mk V and Mk IX before it, the Mk XII was introduced as an interim variant while Supermarine continued to develop the optimum Griffon-powered variant featuring a highly modified wing. With its single-stage supercharger, the Griffon II engine conferred sparkling performance at a low level on the Mk XII, but at higher levels, the aircraft was notably inferior to

SPITFIRE MK IV (LATER MK XX) DP845, the initial Griffon Spitfire prototype, seen here at Boscombe Down, was the aircraft used in the race with the Typhoon and Fw 190. As originally built it featured a standard Mk V rudder, standard wingtips and a four-blade propeller and spinner as fitted to the Mk IX.

the Mk IX. As such, the usage of the Mk XII was limited to low- and medium-altitude operations, and all production Mk XIIs featured the low-level-specific clipped wing. In service, pilots soon discovered that the throttle had to be handled extremely carefully on take-off, as the enormous torque generated by the Griffon could easily develop into an uncontrollable swing, even despite the enlarged rudder. This tendency could be compounded by the reversal of propeller

Mks XII–XVIII: Griffon Engine Variants

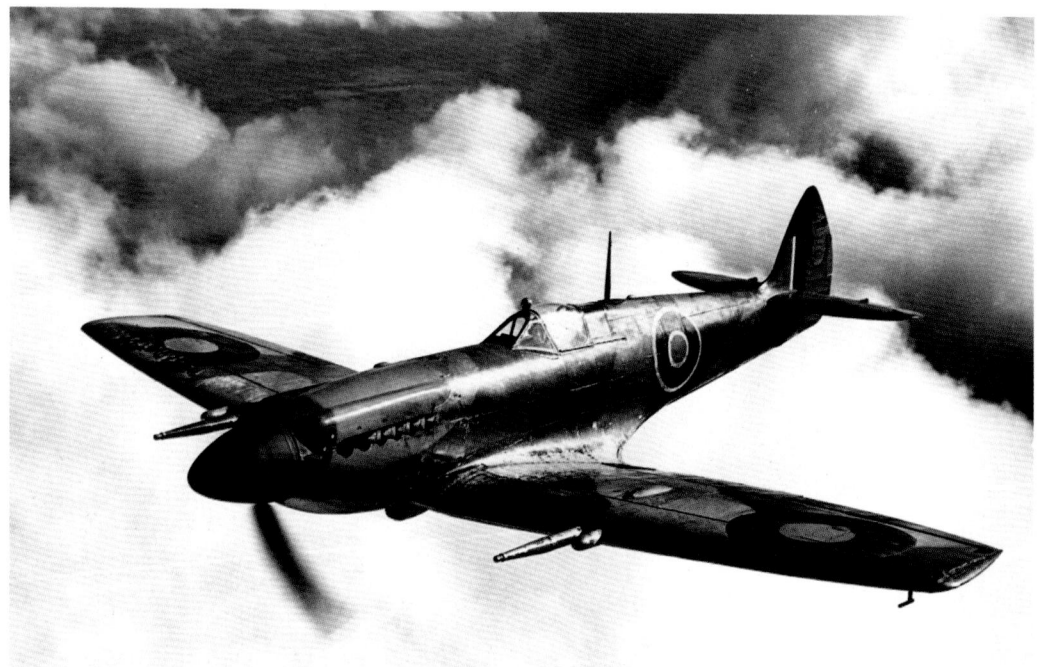

rotation in the Griffon: pilots used to Merlin Spitfires were accustomed to counteracting a swing to port, and the appearance of a more pronounced swing to starboard could catch out a pilot unfamiliar with the type with potentially serious results. Only two squadrons were ever equipped with the Mk XII: the first, 41 Squadron, became operational with the new aircraft in April 1943 and scored the first victory by a Griffon Spitfire on 17 April. German fighters were reportedly reluctant to engage with any Spitfire below 6096m (20,000ft), so opportunities for air-to-air combat were relatively rare – but on the occasions that Luftwaffe fighters could be drawn into a dogfight, the Mk XII proved

formidable. After a busy period escorting bombers over occupied France and providing air support to the D-Day landings, the Mk XIIs found themselves in high demand as one of the few types able to successfully intercept the V-1 missile. 82.5 V-1s were credited to Mk XII pilots before surviving aircraft were retired in September 1944. By this time, a new and much more versatile Griffon-powered Spitfire variant had entered service: the Mk XIV.

Development of the Mk XIV

By 1943, Rolls-Royce had developed a Griffon with two-stage supercharging, allowing it to maintain power output at much higher altitudes. This was the first of the Griffon 60 series, so

PROTOTYPE TRIALS
DP845, the original Griffon Spitfire was modified to production standard as the prototype Mk XII. Photographed during trials in September 1942, the aircraft now featured a broad chord pointed rudder, cannon armament, clipped wings and a much larger propeller spinner.

named because it aligned with the highly successful Merlin 60 series fitted with two-speed, two-stage superchargers. Promising power in the 2000–2300hp class, the new Griffon offered a significant leap in rated output over the Griffon II as fitted to the Spitfire Mk XII. As a result, Supermarine began undertaking a major redesign of the airframe to best take advantage of this extra power.

Mks XII–XVIII: GRIFFON ENGINE VARIANTS

This aircraft would eventually emerge in the form of the Spitfire Mk 21, but in the interim, six Mk VIII airframes were delivered to Rolls-Royce to serve as test beds for the new Griffons. Flight testing quickly revealed that the Mk VIII/Griffon 60 combination possessed excellent performance: sufficient to justify putting this interim aircraft into immediate production. This new variant was expected to be in service considerably earlier than the Mk 21, and this would prove to be the case, as Mk 21 development ran into serious delays. The Mk XIV, as the Mk VIII/Griffon conversion was designated in production form, would become the last variant to be produced in large numbers, with 957 examples eventually rolling off the assembly line. Once again, a more developed Spitfire variant was sidelined by a largely ad hoc conversion, and the Spitfire Mk 21 would be produced only in trivial numbers (though this likely would not have been the case had hostilities continued).

In production form, the Mk XIV was equipped with the Griffon 65 engine delivering 2035hp and utilizing a five-bladed Rotol propeller. The two-stage supercharger needed its own intercooler radiator, just like its Merlin counterpart, and so the Mk XIV featured symmetrical radiators under each wing, as fitted to the Mk IX and VIII, with engine coolant and oil cooler mounted under the port wing, and intercooler and a second engine coolant radiator to starboard. Like the Mk VIII and later production Mk IXs, the air intake featured a tropical filter from the factory, and all Mk XIVs were fitted with a retractable tailwheel. The large supercharger installation on the Griffon 60 series necessitated lengthening the nose once again, thus requiring the introduction of a further enlarged fin and rudder, though horizontal tail surfaces were the same as those fitted to later Merlin-engine marks.

The first Mk XIVs were fitted with the C-type wing, but most Mk XIVs featured the E-type wing with two 20mm (0.79in) Hispano and two 12.7mm (0.50in) Brownings. Later production XIVs were built with the cut-

SPITFIRE MK XII

The Mk XII was a low-altitude aircraft and as such too specialized to see widespread use. Only two squadrons ever flew the Mk XII, this example is from the initial unit to convert to the type, 41 Squadron.

SPITFIRE Mk XII

Weight (maximum take-off): 3363kg (7415lb)
Dimensions: Length 9.62m (31ft 7in), Wingspan 9.9m (32ft 6in), Height 3.86m (12ft 8in)
Powerplant: One 1293kW (1735hp) Rolls-Royce Griffon III liquid cooled V-12 piston engine
Maximum speed: 639km/h (397mph)
Range: 793km (493 miles)
Ceiling: 11,887m (39,000ft)
Crew: 1
Armament: Two 20mm (0.79in) Hispano cannon and four 7.7mm (0.303in) Browning machine guns in wings; up to 460kg (1000lb) bomb load

Mks XII–XVIII: Griffon Engine Variants

SPITFIRE F MK XIV
The Squadron Leader's 'flash' under the windscreen signifies that this aircraft of 610 Squadron belongs to the commanding officer, Richard Newbury. During September 1944, 610 Squadron was engaged in 'Anti-Diver' patrols from Lympne in Kent.

SPITFIRE Mk XIV

Weight (maximum take-off): 3856kg (8500lb)
Dimensions: Length 9.96m (32ft 8in), Wingspan 11.23m (36ft 10in), Height 3.86m (12ft 8in)
Powerplant: One 1508kW (2050hp) Rolls-Royce Griffon 65 liquid cooled V-12 piston engine
Maximum speed: 721km/h (448mph)
Range: 740km (460 miles)
Ceiling: 13,564m (44,500ft)
Crew: 1
Armament: two 20mm (0.79in) Hispano cannon and four 7.7mm (0.303in) or two 12.7mm (0.5in) Browning machine guns in wings; up to 460kg (1000lb) bomb load

down rear fuselage, and many Mk XIVs were modified in service to feature clipped wings for low-level operations.

Included within the total of Mk XIVs built was a considerable number of FR Mk XIVEs, which retained the same armament as the Mk XIVE but featured an F24 camera mounted obliquely in the rear fuselage immediately behind the cockpit, as well as an extra 150L (33 gallon) fuel tank in the rear fuselage. All FR Mk XIVs were delivered with the cut-down rear fuselage and were fitted with a larger rudder, greater in chord and height, to deal with the resultant reduction in directional stability.

Combat debut

Deliveries of the Mk XIV began from Supermarine in October 1943 and the new Spitfire entered service with 610 Squadron in January 1944, entering operations in May. In the meantime, a standard production Spitfire Mk XIV had been tested at the Air Fighting Development Unit (AFDU) in comparative tactical trials against the Tempest Mk V, Mustang Mk III and Spitfire Mk IX, as well as in combat trials against captured examples of the Fw 190A and Me 109G. The results were exceptionally good, with the Spitfire being described as 'superior to the Me.109G (sic) in every respect' and proving better than the Fw 190 in all areas except rate of roll.

When judged against the other Allied types, the Mk XIV was superior to the Mk IX in all respects, as one might expect. The Tempest possessed a slight edge at low levels but was outperformed by the Mk XIV above 6096m (20,000ft) or so. Against the Mustang, there was little to choose between the two types, except in range, where the Mustang enjoyed a huge advantage. The AFDU report hedged its bets rather by stating: 'With the exception of endurance no conclusions can be drawn, as these two aircraft should never be enemies. The choice is a matter of taste.'

By this juncture, the basic airframe design of the Spitfire was (in wartime terms at least) getting a little long in the tooth, so it is all the more impressive to note that the AFDU concluded that the

Mks XII–XVIII: GRIFFON ENGINE VARIANTS

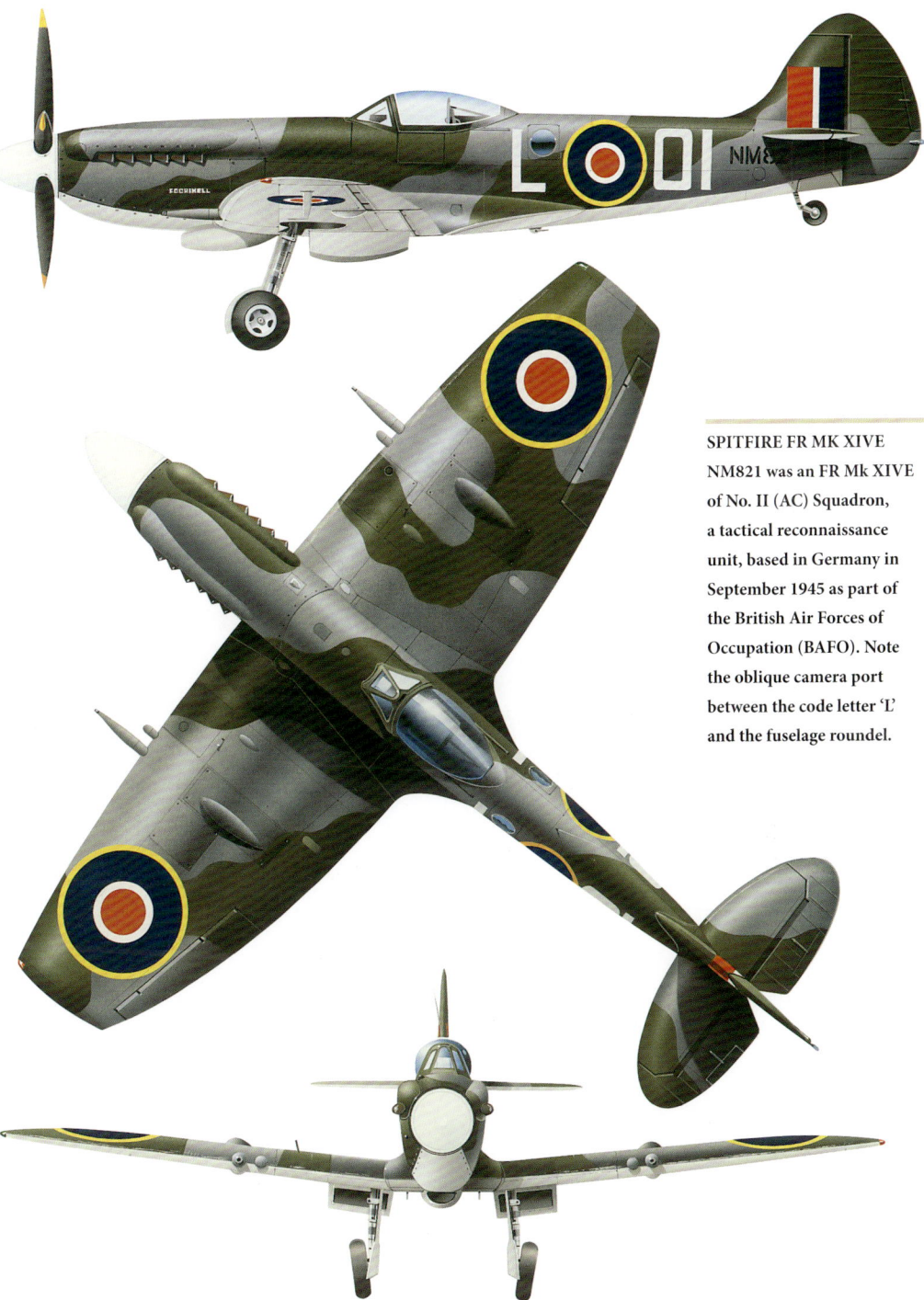

SPITFIRE FR MK XIVE NM821 was an FR Mk XIVE of No. II (AC) Squadron, a tactical reconnaissance unit, based in Germany in September 1945 as part of the British Air Forces of Occupation (BAFO). Note the oblique camera port between the code letter 'L' and the fuselage roundel.

Mks XII–XVIII: Griffon Engine Variants

Spitfire Mk XIV 'has the best all-round performance of any present-day fighter, apart from range'.

From mid-June 1944, 610 Squadron was in on 'Anti-Diver' patrols: the RAF's defence against V-1 flying bombs that had begun to be launched at Britain one week after the Allied D-Day landings in Normandy. 601 was soon followed by a further six squadrons, including one Dutch, one Belgian and an RCAF unit. The Mk XIV, primarily due to its high top-speed, proved well suited to this role, and Spitfire Mk XIVs were credited with the destruction of 303 V-1s, making them the third most successful aircraft type against the flying bombs after the Tempest V and Mosquito.

Attacking these missiles was a hazardous undertaking. The V-1 cruised at a relatively low level, close to the maximum speed of the Spitfire Mk XIV, and an attacking aircraft could (and frequently did) detonate the warhead with cannon fire. If this occurred, at the speeds these actions were taking place, there was no time for the pursuing aircraft to avoid the explosion, which could result in damage to, or even the destruction of, the attacker. Pilots quickly learned that the slipstream of their aircraft, if flown directly in the flight path of the V-1, could upset the sensitive controls of its autopilot, causing it to crash.

Some pilots even resorted to 'tipping' the missile: formatting with the V-1 and using their wingtip to lift the wing of the flying bomb, sending it out of control. During this period, offensive patrols over the Channel were also flown, and following the advance of Allied armies into France and beyond, most of the squadrons equipped with Mk XIVs were operating in continental Europe with 2TAF. With the threat from enemy air power rapidly diminishing, many Mk XIVs, like their Mk IX and VIII forebears, were pushed into the fighter-bomber role. As the war in Europe came to its conclusion, efforts were undertaken to re-equip squadrons in the Far East with the Mk XIV for operations against Japan, but although two squadrons received Mk XIV aircraft before VJ Day, neither squadron had attained operational status before the conflict with Japan ended.

Mk XVIII

Following the Mk XIV on the production line came the Spitfire Mk XVIII. Externally more or

SPITFIRE FR MK XVIII
Sporting Korean War identification stripes, this Spitfire FR XVIII of 28 Squadron was based at Kai Tak, Hong Kong, in 1950. 28 Squadron was the last frontline RAF fighter squadron to fly the Spitfire, converting to the Vampire FB.5 in February 1951.

Mks XII–XVIII: GRIFFON ENGINE VARIANTS

less indistinguishable from its immediate predecessor, the Mk XVIII featured a marginally longer fuselage and a strengthened wing and undercarriage. Internal fuel capacity was usefully increased with the addition of two 150L (33 gallon) tanks in the rear fuselage instead of the single 131.84L (29 gallon) tank of the FR Mk XIV. All Mk XVIIIs featured the low-back cut-down rear fuselage, teardrop canopy and the enlarged tail surfaces of the FR Mk XIV.

Postwar service
Production of the new mark totalled 300, of which 100 were of the FR Mk XVIII variant, with the option for two vertical and one oblique F.24 cameras, or a single vertical F.52, in the rear fuselage. Deliveries of the Mk XVIII started in mid-1945 and, as a result, the aircraft did not see service during the war, but did serve after it, with six squadrons in the Middle and Far East.

SPITFIRE FR Mk XVIII
Weight (maximum take-off): 4218kg (9300lb)
Dimensions: Length 9.96m (32ft 8in), Wingspan 11.23m (36ft 10in), Height 3.86m (12ft 8in)
Powerplant: One 1721kW (2340hp) Rolls-Royce Griffon 67 liquid cooled V-12 piston engine
Maximum speed: 703km/h (437mph)
Range: 980km (610 miles)
Ceiling: 13,564m (44,500ft)
Crew: 1
Armament: two 20mm (0.79in) Hispano cannon and four 7.7mm (0.303in) or two 12.7mm (0.5in) Browning machine guns in wings; up to 460kg (1000lb) bomb load or up to eight 27kg (60lb) RP3 rockets under wings

Photo Reconnaissance Models

In April 1941, Squadron Leader Tony Martindale, an experienced test pilot at the A&AEE, initiated a full-throttle 45-degree dive in a Spitfire from 12,192m (40,000ft). As he began to consider recovering from the dive, 'there was a fearful explosion and the aircraft was enveloped in white smoke'. The propeller and reduction gear had torn themselves off the aircraft, one of the Merlin's connecting rods had punched its way through the crankcase and the entire engine had shifted sideways. With the weight of the propeller and gearbox now removed, the aircraft became tail-heavy, and the dive developed into a zoom climb, this uncommanded manoeuvre subjecting the aircraft to an 11G load and causing Martindale to black out. After a few seconds, although unable to see out of the windscreen and most of the canopy due to it being covered in oil from the now-departed reduction gearing, Martindale was able to ascertain that he was still climbing and found that he could open the canopy. Preparing to bail out, he then discovered that he still had the aircraft under control and elected instead to glide the 32.2km (20 miles) back to Farnborough, where he pulled off a perfect three-point landing.

UNARMED AND ALONE
Most PR Spitfire reconnaissance flights were solitary affairs, with the unarmed aircraft penetrating deep into enemy airspace. This is EN654, the prototype PR MK XI, here flown by Jeffrey Quill, Supermarine's chief test pilot.

PHOTO RECONNAISSANCE MODELS

On landing, it was found that in addition to the other damage his Spitfire had suffered, the wings of the aircraft had been swept back slightly by the stress of pulling out from the dive. What had not been damaged, however, was the camera that had been recording Martindale's instruments for the duration of his eventful flight, and this had recorded that he had attained a true airspeed of 620 mph, or Mach 0.92, the highest known speed attained by any propeller-driven aircraft during the war. Yet this – the fastest Spitfire ever – was not a fighter at all, but an unarmed reconnaissance aircraft.

Reconnaissance role

Reconnaissance in the years leading up to the war was generally considered to require a relatively large, two-seat aircraft. Aerial cameras were bulky and needed a second crew member to operate them, and for long-range strategic reconnaissance missions, a large fuel load was required. As a result, the RAF's standard long-range reconnaissance aircraft at the outbreak of war was the Bristol Blenheim, a twin-engine light bomber. Short-range reconnaissance was handled by the Westland Lysander, an Army Co-operation aircraft that possessed excellent STOL capability, which

HIGH-ALTITUDE VARIANT
Operating in the high-altitude photo-reconnaissance role, the PR.Mk XIXs were the last variants to see active service with the RAF. It was not until July 1957 that the one remaining example, PS853, was finally retired.

would later see it used with great success for the nocturnal insertion and recovery of agents into occupied France. However, the Westland Lysander lacked the performance needed to evade enemy fighters by day. The early war years saw the vulnerability, and heavy losses, of both types rapidly become apparent when they were

PHOTO RECONNAISSANCE MODELS

committed to reconnaissance missions. An alternative was urgently required. The radical decision to use a fast single-seat fighter for the reconnaissance mission was largely the result of the work of flamboyant Australian aerial photography pioneer Sidney Cotton, commander of 1 RAF Photographic Development Unit, and his assistant, Flying Officer Maurice 'Shorty' Longbottom. This partnership was to revolutionize British aerial reconnaissance

SPITFIRE PR MK TYPE IA

Two Spitfire Mk IA fighters (N3069 and N3071) were converted to unarmed 'Type A' standard, with a 12.7cm (5in) focal length, vertically-mounted camera in each wing and a coat of light-green paint known as 'Camoutint'. Here, N3071 is seen at Séclin, France, on 18 November 1939, being run up prior to making the historic first PR flight.

and would prove both hugely successful and highly influential, not least in the development of the Spitfire. Cotton believed that a high-speed, high-altitude reconnaissance platform was required, but it was Longbottom who produced a memorandum informed by Cotton's ideas entitled 'Photographic Reconnaissance of Enemy Territory in War', which he submitted to the Air Ministry in 1939. In this document, Longbottom suggested that in reality, strategic reconnaissance work is more dangerous than bombing, and proposed a solution: 'Reconnaissance must be done in such a manner as to avoid the enemy fighter and defences as completely as possible. The best method of doing this appears to be the use of a single small machine relying on its speed, climb and ceiling to avoid detection.'

'Heston Flight'

After the outbreak of war, Cotton was commissioned into the RAF as a squadron leader and placed in command of the highly secret 'Heston Flight' charged with performing strategic reconnaissance. Initially, Cotton's own privately owned Lockheed 12A was utilized, but both he and Longbottom felt that a suitably modified Spitfire would be far more effective. Luckily for the future operations of the Heston Flight, it came under the aegis of Fighter Command, and whilst visiting Heston in October 1939, Air Chief Marshal Hugh Dowding was persuaded by Cotton to provide the flight with a pair of Spitfires. This was surprising, since at the time the Spitfire was both in great demand as a fighter and in terrifically short supply. The two Spitfire Mk Is arrived at Heston

PHOTO RECONNAISSANCE MODELS

the day following Dowding's visit, and modifications began to equip them for their new role, focusing particularly on improving the speed and range of the aircraft as well as fitting appropriate cameras within them, a process jocularly referred to as 'Cottonizing'.

Blister canopies

These first two Spitfires had all armament and radio equipment removed. In each Spitfire, the empty gun ports and all panel lines were filled with plaster of Paris, while the entire airframe was polished to give a highly smooth surface. The aircraft were finished in a pale blue-green shade called Camoutint. It was developed by Cotton after observing an aircraft painted this colour disappearing from view against the sky very quickly despite visibility being very good. The Spitfires each carried two F24 cameras with a 12.7cm (5in) focal length lens mounted within the wings, one on each side pointing directly downwards, in the space previously occupied by the inboard wing guns. Heating equipment was fitted to prevent the cameras from freezing and frost forming on the lens at high altitude. The decreased weight and exceptional surface finish conferred upon the modified Spitfires a maximum speed of 627.6km/h (390mph) – around 48km/h (30mph) faster than the standard Mk I. All conversion work at this point was undertaken by the Heston Aircraft Company, with no input from Supermarine, and these first two reconnaissance Spitfires were retrospectively designated the Mk I PR Type A. One feature of most reconnaissance Spitfires was the specially 'blown' canopies incorporating large lateral teardrop blisters, allowing pilots a considerably clearer view to the rear and below, which was needed to sight the cameras. On all the unarmed PR conversions, the Spitfire's gunsight was swapped for a small camera control box that was used to turn the cameras on, control the time intervals between photos and set the number of exposures.

First mission

On 18 November 1939, the first RAF high-speed, high-altitude photo-reconnaissance mission of the war, and indeed the first Spitfire flight over enemy-held territory, was flown by Longbottom himself. Flying from Seclin, in France, he attempted to photograph Aachen from 10,085m (33,000ft), but bad weather forced him to return. Later sorties were more successful: in December, Longbottom took some excellent photos of the Siegfried line, vindicating the PR Spitfire concept. Further missions saw the Spitfire photograph Cologne,

---

SPITFIRE PR Type 1A

Weight (maximum take-off): 3164kg (6975lb)

Dimensions: Length 9.12m (29ft 11in), Wingspan 11.23m (36ft 10in), Height 3.02m (9ft 10in)

Powerplant: One 770kW (1030hp) Rolls-Royce Merlin III liquid cooled V-12 piston engine

Maximum speed: 628km/h (390mph)

Range: 1207km (750 miles)

Ceiling: 10,670m (35000ft)

Crew: 1

Armament: N/A

SPITFIRE PR TYPE 1A

Their unique colour schemes made the early PR Spitfires distinctly striking machines. N3071 was one of the initial two Spitfires converted at Heston for the PR role.

PHOTO RECONNAISSANCE MODELS

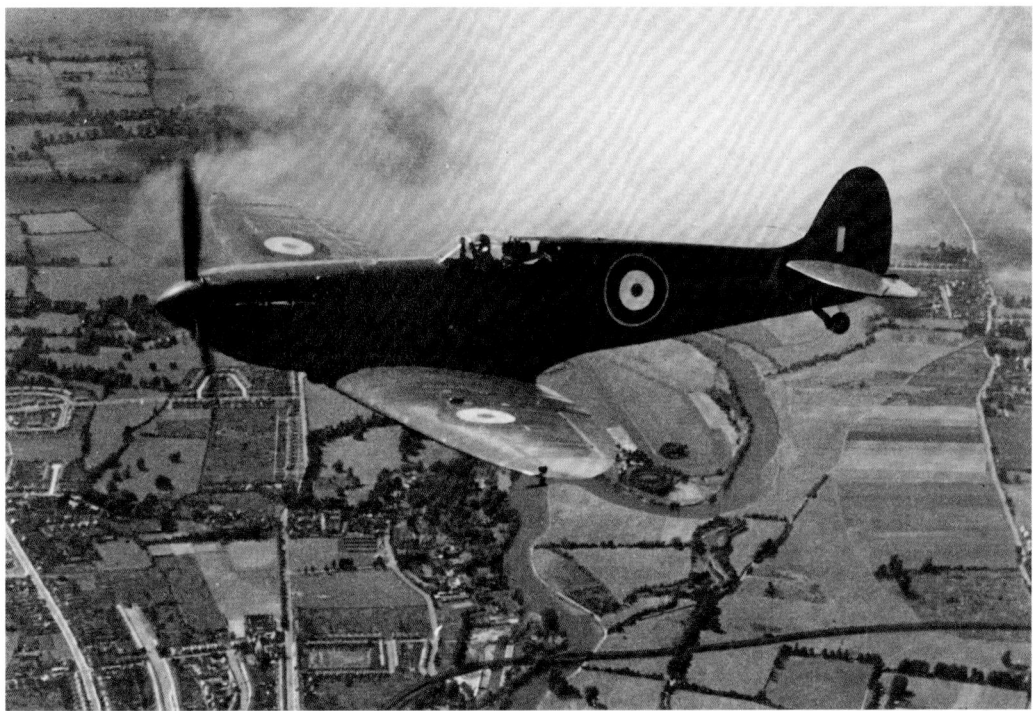

Kaiserslautern, Wiesbaden, Mainz and the industrial heartland of the Ruhr. No fighter version of the Spitfire would operate from France before 1944, and the only Spitfires to fly in France before the German occupation were the reconnaissance aircraft of Cotton's Heston Flight, which had meanwhile been renamed 'No. 2 Camouflage Unit' to disguise its true purpose. However, with the increasing use of the PR Spitfire, on 17 January 1940, No. 2 Camouflage Unit was renamed the 'Photographic Development Unit' (PDU), while another PR Spitfire unit, 212 Squadron, was formed in France. During the early stages of its existence, the PDU would be known as 'Cotton's Club', or, less flatteringly, as 'Cotton's Crooks' (mainly due to Cotton's tendency to ignore regulations).

MR type

Initial missions demonstrated that the concept of using the Spitfire as a reconnaissance aircraft was sound but that greater range was required. In the Mk I PR Type B (also known as 'Medium Range' or 'MR') conversions that followed, an extra 131.84L (29 gallon) fuel tank was installed in the rear fuselage. At the time, this was the only extra tankage the RAE would approve. The F24 camera lenses were replaced with a 20.3cm (8in) focal length, resulting in images up to a third larger in scale. Much larger cameras could be installed

SPITFIRE MK I PR TYPE C
The Spitfire PR Mk IC introduced a further increase in range over the Mk IB, by adding a 136-litre (30-imp gal) fuel tank under the port wing. This counterbalanced the two 20.3cm (8in) lens cameras in the flattened blister under the starboard wing.

in the fuselage immediately behind the pilot, but it was believed this would upset the Spitfire's centre of gravity. Cotton demonstrated that by removing the lead weights installed in the very end of the rear fuselage to balance the weight of the constant speed propeller units, it would be possible to install cameras with an even longer focal-length lens in the Spitfire's fuselage. The Type B also dispensed with

PHOTO RECONNAISSANCE MODELS

the heavy bulletproof windscreen for the first time and introduced a colour change, being painted overall blue-grey in a shade later to be known as PRU blue (from 'Photographic Reconnaissance Unit', as the PDU was renamed in mid-1940). Several other shades would be utilized before PRU blue became the standard colour scheme for photo reconnaissance aircraft.

Mk I PR Type C

Further range was delivered by the Mk I PR Type C (also known as the 'Long Range' or 'LR' Spitfire), which carried a total of 654.64L (144 gallons) of fuel and could fly as far as Kiel. The extra fuel was carried in the 131.84L (29 gallon) tank behind the pilot and in a 136.38L (30 gallon) blister tank under the port wing, which was counterbalanced by a fairing for the camera under the starboard wing. A larger oil tank was installed, which necessitated the reshaping of the nose contours below and behind the propeller spinner, resulting in the distinctive PR Spitfire nose profile.

As well as its sterling work as a reconnaissance platform, a PR Type C managed to score an air-to-air 'kill' despite being completely unarmed. On 13 June 1940, Flying Officer George Patterson Christie, a Canadian pilot of 212 Squadron who would later fly fighter Spitfires during the Battle of Britain, repeatedly dived at an Italian Fiat BR.20 bomber off the coast of Monaco, forcing it to ditch in the sea. Awarded the Distinguished Flying Cross for this action, he was also berated by Cotton for engaging in combat

KEY ROLE

Guy Gibson, leader of the famed 'Dam Busters' raid points out details of a post-strike photograph of the breached Moehne dam obtained by a PR Spitfire. Spitfires had already supplied reconnaissance photographs with which the raid was planned.

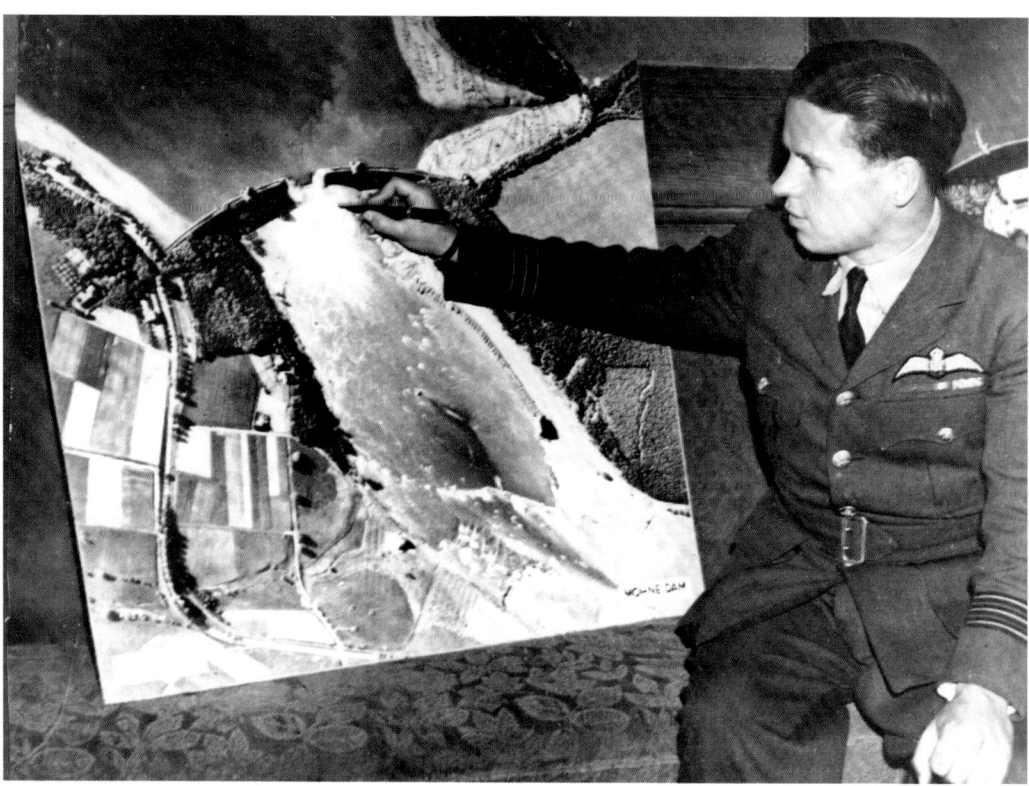

MK I PR TYPE E

Before the Type D entered production, the single Mk I PR Type E had appeared, of which only a single example was built. A need was identified for an aircraft capable of taking oblique close-up photographs as opposed to the high-altitude vertical pictures that had been delivered so far by PR Spitfires.

The Type E conversion mounted a forward-facing F.24 camera in a fairing under each wing: each camera pointed slightly outwards and was angled downwards 15 degrees. The lateral cameras were aimed by lining up a small + symbol marked on the side of the canopy, with a thin black line painted onto the aileron. Operating when weather conditions precluded high-altitude reconnaissance, the Type E's sorties could prove dramatic, such as a mission on 7 July 1940, when the Type E photographed Boulogne from a height of approximately 91.4m (300ft). As this sort of operation was considered somewhat hazardous, the Type E was generally only used in conditions of at least 85 per cent cloud cover. The PR Type F, meanwhile, was an interim 'super-long-range' conversion prepared by Heston Aircraft, pending delivery of the factory-built Type D. The Type F carried a 136.38L (30 gallon) fuel tank under each wing in addition to the 131.84L (29 gallon) tank in the rear fuselage and the enlarged oil tank in the nose. It proved such a significant improvement that nearly all existing Type Bs and Type Cs were re-converted to Type Fs. The range capability was impressive – flying from bases in East Anglia, the PR Type F could photograph Berlin and return, the first such trip being made on 14 March 1941.

SINGLE EXAMPLE
This unique PR Mk IV was modified by 103 MU at Aboukir in Egypt with extended wingtips and a tuned Merlin engine. The unusual patchy finish was caused by the application of filler over surface irregularities that was then polished to perfect smoothness. The aircraft was used by 680 Squadron in 1943 for sorties over Salonika.

PHOTO RECONNAISSANCE MODELS

when his primary function was to return with his reconnaissance photographs. However, a few days later, Cotton himself would be removed from his post after accepting a fee to fly Marcel Broussac, the hugely wealthy owner of the Christian Dior fashion house, out of occupied France, and would play no further role in the operations of the Photographic Reconnaissance Unit, which he had been instrumental in creating.

Mk I PR Type D

The Mk I PR Type D (or the 'Extra Super Long-Range Spitfire') was the first PR variant that was not a conversion of existing fighters, and the first to be built in large numbers. Produced by Supermarine at Woolston, the contract was placed at a moment when fighter production was desperately needed, and the first two factory-built reconnaissance Spitfires only appeared in October, sometime after the later Type E and F conversions were completed at Heston. The new version introduced integral tanks in the D-shaped wing leading edges, ahead of the main spar, resulting in a so-called 'wet wing' with 259.13L (57 gallons) of fuel in each wing. The first two aircraft retained the 131.84L (29 gallon) tank in the rear fuselage and an additional 63.65L (14 gallon) oil tank was fitted in the port wing. The cameras were two vertically mounted F24s with either a 20.3cm (8in) or 50.8cm (20in) lens, or two vertically mounted F8s with 20in lens, mounted in the rear fuselage. When fully fuelled, the centre of gravity was so far back that the aircraft was difficult to control until the rear fuselage tank had emptied. Despite this, the type quickly demonstrated

SPITFIRE PR TYPE G (LATER PR MK VII)
Finished in an overall light pink shade developed to conceal the aircraft against a layer of cloud, R7059 was photographed in flight over the UK during 1941. This aircraft is depicted in the three-view overleaf.

its impressive capability, photographing such distant targets as Stettin, Marseilles, Trondheim and Toulon.

PR Mk IV

With the concept successfully proven by the first two aircraft, the next Type D Spitfires, soon renamed PR Mk IV, were modified by deleting the rear fuselage fuel tank and increasing the capacity of the leading-edge tanks to 302.32L (66.5 gallons) each. In this form, the PR Mk IV proved to be a much easier-to-fly aircraft, and the pilot's

PHOTO RECONNAISSANCE MODELS

well-being was further enhanced by the provision of cockpit heating for the first time, a welcome addition in an aircraft that was flying missions of up to seven hours in sub-zero temperatures.

Ultimately, a total of 229 PR Mk IVs were built, reflecting the increasing importance of the Spitfire in the reconnaissance role. The PR Mk IV was also the first reconnaissance Spitfire to see service with another nation. Two examples were supplied to the Soviet Union in 1942 and were intensively used for the remainder of the war, carrying out many missions over Norway and photographing the *Tirpitz* at anchor.

PR Type G

The successful use of the solitary Type E directly informed the development of the PR Type G, which was the first fighter-reconnaissance version and was used for a similar low-level tactical role. Fully armed with eight 7.7mm (0.303in) Brownings and retaining the armoured windscreen and gunsight, the new version was able to defend itself if intercepted. One oblique F24 camera, with either a 20.3cm (8in) or 35.6cm (14in) lens, was fitted in the fuselage along with two vertical F24s. The first Type Gs were converted from Mk I airframes re-engined with Merlin 45s replacing the Merlin IIs as originally fitted. Later PR G conversions were based on Mk V airframes. Heston Aircraft converted 45 Spitfires into PR type

Gs. These aircraft would become the PR Mk VII when a new system of mark numbers for PR Spitfires was introduced in 1941. Of the other variants still in service, the Type C became the PR Mk III, and the Type D, as previously noted, became the PR Mk IV. The single Type E instead became the only PR Mk V, and the Type F became the PR Mk VI.

PR Mk IX

With the introduction of the Merlin 60 series to produce the Mk IX fighter, it wasn't long before a PR version was developed. By this time, reconnaissance Spitfires were correctly regarded as an invaluable intelligence-gathering asset, and their development and production were to take place in-house at Supermarine. A dedicated PR variant utilizing the Merlin 60 series combined with the best aspects of the reconnaissance Spitfires built so far was under development, but as usual with Spitfire evolution, a somewhat ad-hoc conversion was rapidly produced before the more sophisticated machine was ready. The PR Mk IX (there being no PR Mk VIII) was the result of taking three Mk IX fighters off the production line and modifying them to fit two vertical F24 cameras in the rear fuselage. The first was delivered to 515 Squadron in November 1942, and following successful initial use, a further 15 Mk IXs were converted. The PR Mk IX featured the distinctive wrap-around PR windscreen and

the deeper nose with the larger oil tank. All armament was removed and a PRU blue finish applied, but the Mk IX conversion did not feature the 'wet wing' integral fuel tanks of the earlier PR Mk IV, achieving adequate range capability by using external drop tanks. Despite only a small number of PR Mk IX conversions being made, the aircraft were involved in some of the most famous reconnaissance missions of the war: a PR Mk IX photographed the target dams in preparation for Operation Chastise, the 'Dambusters' mission, and brought back pictures of the breached Moehne Dam on the day after the attack. Confusingly, a different Mk IX conversion resulted in the FR Mk IX (FR standing for Fighter Reconnaissance). This aircraft retained the full armament of the Mk IX but featured a single, port-facing, oblique camera. These aircraft were used for low-level tactical reconnaissance missions (known as 'dicing') in support of operations by the Army. 16 Squadron utilized FR Mk IXs, painted pale pink to camouflage the aircraft against cloud cover, to reconnoitre the Arnhem area before the controversial Operation Market Garden.

PR Mk X

The next numerical PR variant, the PR Mk X, was originally intended to be designated the PR Mk VII, as it was a dedicated reconnaissance variant of the pressurized Mk VII fighter. However, a policy change saw it

PHOTO RECONNAISSANCE MODELS

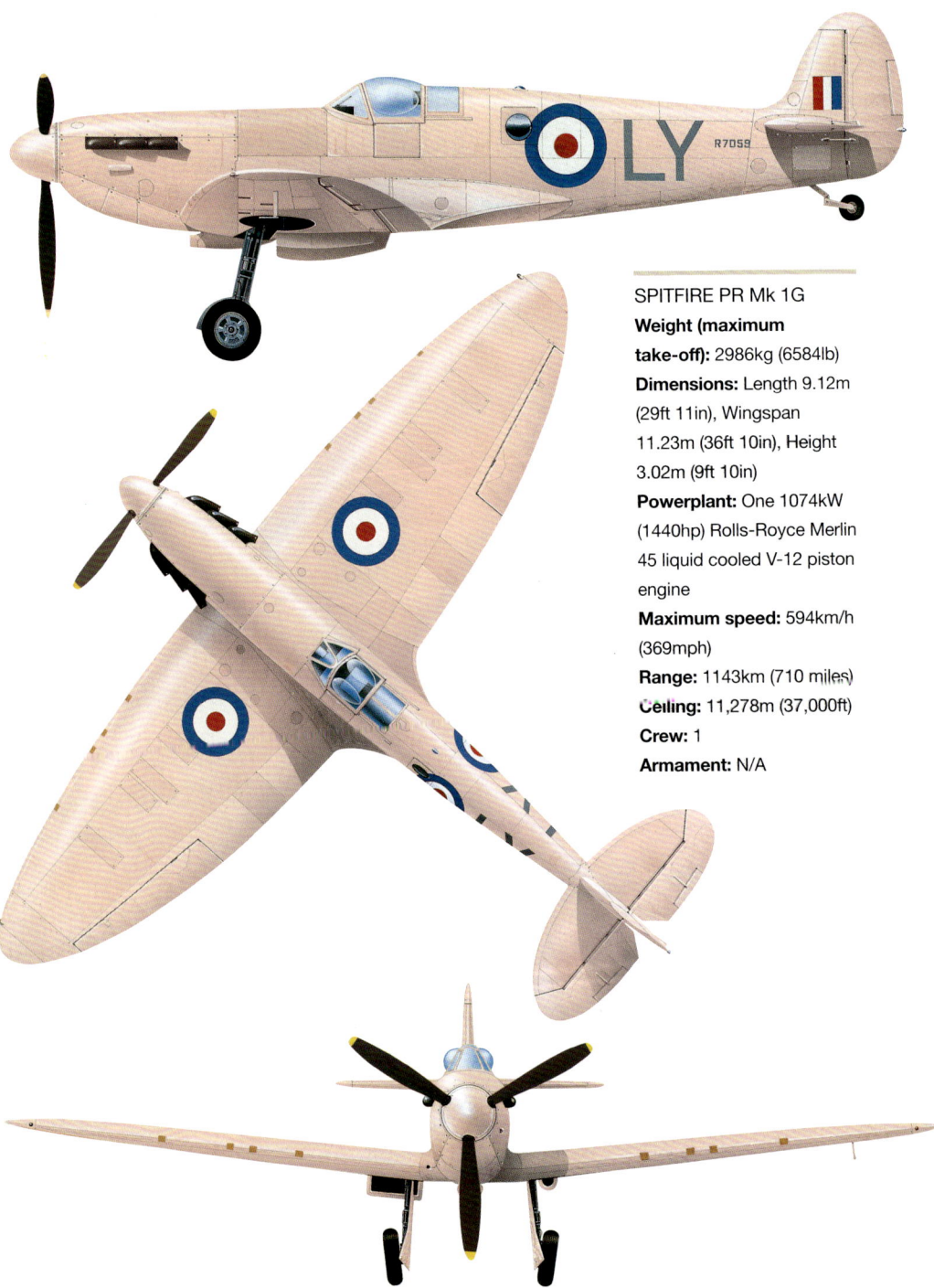

SPITFIRE PR Mk 1G
Weight (maximum take-off): 2986kg (6584lb)
Dimensions: Length 9.12m (29ft 11in), Wingspan 11.23m (36ft 10in), Height 3.02m (9ft 10in)
Powerplant: One 1074kW (1440hp) Rolls-Royce Merlin 45 liquid cooled V-12 piston engine
Maximum speed: 594km/h (369mph)
Range: 1143km (710 miles)
Ceiling: 11,278m (37,000ft)
Crew: 1
Armament: N/A

PHOTO RECONNAISSANCE MODELS

redesignated the PR Mk X before it began appearing. Adding to the muddled development of the PR Spitfires, the Mk X actually followed the PR Mk XI into service. In addition, a PR Mk VIII, based on the Mk VIII airframe, was also planned, with an order for 70 placed in April 1942, but these would never actually be built, the aircraft in question emerging as standard fighter Mk VIIIs instead. Sixteen examples of the PR Mk X were constructed, and as the first pressurized PR variant, they supplemented the PR Mk XIs of two units: 541 and 542 squadrons.

All 16 saw service at one time or another, and the improved high-altitude comfort afforded by the cabin pressurization system meant that these aircraft could operate at 12,192m (40,000ft) for prolonged periods without the debilitating effects such altitudes would normally have on their pilots. The Mk X played an important role in developing the later Griffon-powered PR Mk XIX, which was also pressurized.

PR Mk XI

In contrast to the scarce PR Mk X, the Mk XI was destined to be built in much greater numbers, with 471 aircraft built before production ended in December 1944. A less specialized and more versatile aircraft than the Mk X, the PR Mk XI was unusual in that it had no direct fighter equivalent utilizing features of the Mks VII, VIII and IX, and was the first PR Spitfire designed as a reconnaissance aircraft from scratch. The Mk XI was the first Spitfire to provide the option of using two F52 cameras with 91.4cm (36in) lenses in the rear fuselage, but a variety of camera fits could be used depending on what the mission dictated. Its performance and capability saw it quickly replace all the Mk I-, II- and V-derived PR conversions then in service. The PR Mk XI's impressive performance also saw it taken on by the USAAF after F-5 photo-reconnaissance versions of the P-38 Lightning twin-engine fighter had proved vulnerable to interception. Despite the relatively high ceiling of the F-5, the USAAF sought an aircraft with still better performance, and the 8th Air Force's 14th Squadron of the 7th Photographic Group acquired several PR Mk XIs, which were nicknamed 'Bluebirds'.

The first USAAF PR sortie was flown on 6 March 1944, when Major Walter L Weitner flew a reconnaissance mission over

Right: D-DAY ROLE
Rolling for the camera is this Mk XI of No. 541 Sqn, RAF. Notable are the 'invasion' stripes (dating the photograph at mid-1944, around the time of Operation Overlord, the Allied invasion of occupied Europe) and the two camera ports in the lower fuselage.

Opposite: SPITFIRE PR MK IG
This aircraft, R7059, was built as a Mk I fighter and as such made its maiden flight from Eastleigh in February 1941. Converted the following month for the low-level reconnaissance role, as a PR Mk IG, the aircraft was delivered to No. 1 PRU in May. Based with a detachment at St. Eval in Cornwall, the aircraft was engaged in low-level 'dicing' sorties over Brest harbour.

101

PHOTO RECONNAISSANCE MODELS

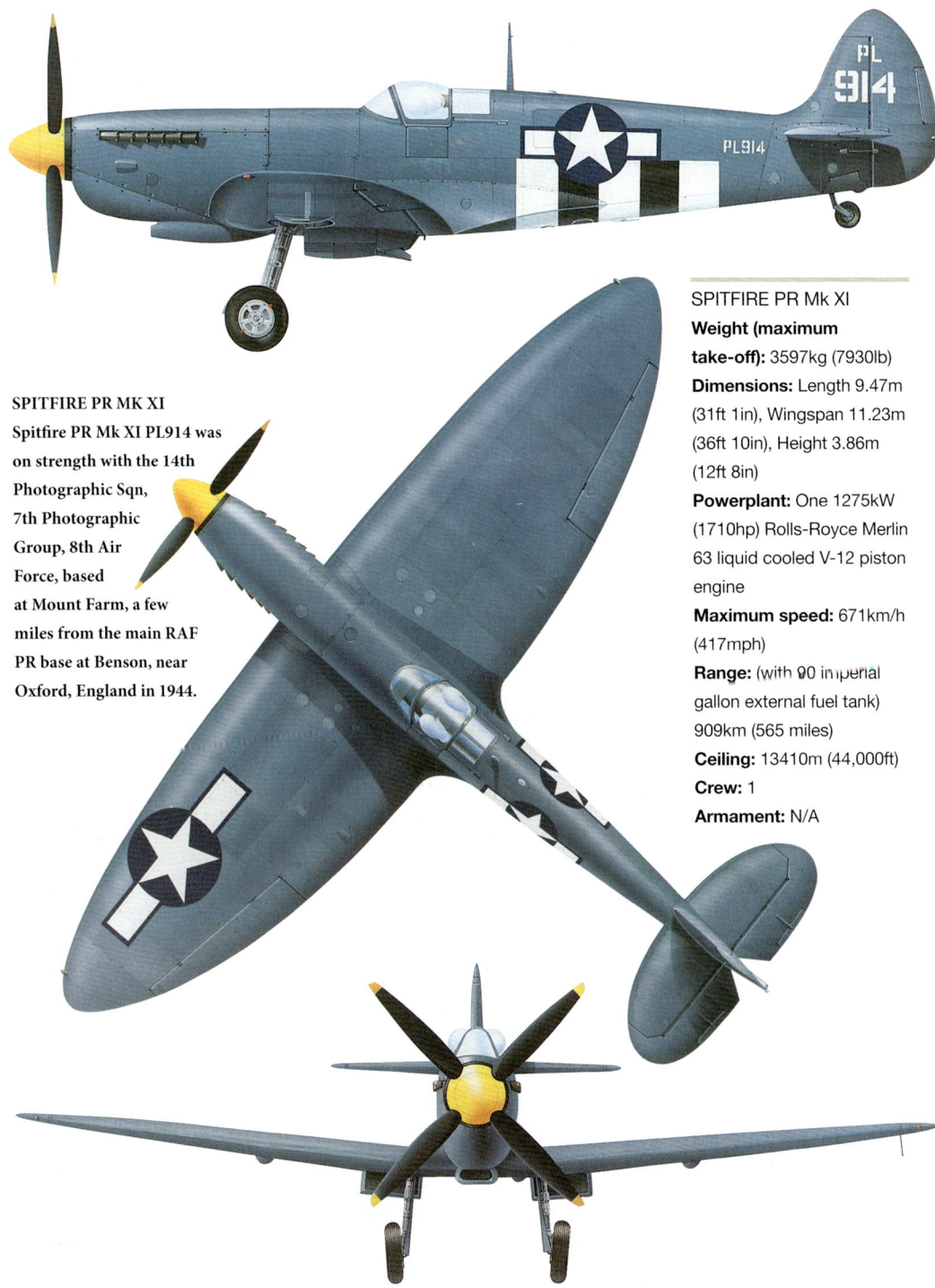

SPITFIRE PR MK XI
Spitfire PR Mk XI PL914 was on strength with the 14th Photographic Sqn, 7th Photographic Group, 8th Air Force, based at Mount Farm, a few miles from the main RAF PR base at Benson, near Oxford, England in 1944.

SPITFIRE PR Mk XI
Weight (maximum take-off): 3597kg (7930lb)
Dimensions: Length 9.47m (31ft 1in), Wingspan 11.23m (36ft 10in), Height 3.86m (12ft 8in)
Powerplant: One 1275kW (1710hp) Rolls-Royce Merlin 63 liquid cooled V-12 piston engine
Maximum speed: 671km/h (417mph)
Range: (with 90 imperial gallon external fuel tank) 909km (565 miles)
Ceiling: 13410m (44,000ft)
Crew: 1
Armament: N/A

PHOTO RECONNAISSANCE MODELS

SPITFIRE PR MK XIX
The PR Mk XIX was the ultimate reconnaissance Spitfire and enjoyed a longer career in British service than any other variant. PS888 made history by performing the last operational RAF Spitfire sortie on 1 April 1954.

SPITFIRE PR Mk XIX
Weight (maximum take-off): 4095kg (9001lb)
Dimensions: Length 9.96m (32ft 8in), Wingspan 11.23m (36ft 10in), Height 3.86m (12ft 8in)
Powerplant: One 1508kW (2050hp) Rolls-Royce Griffon 65 liquid cooled V-12 piston engine
Maximum speed: 717km/h (445mph)
Range: (with 90 imperial gallon external fuel tank) 2896km (1800 miles)
Ceiling: 12,965m (42,600ft)
Crew: 1
Armament: N/A

Berlin. The unit subsequently flew the Spitfire PR Mk XI until the end of the war. One of the unit's few losses occurred on 5 September 1944, when Lt Robert Hillborn was forced to bail out after being intercepted by a Messerschmitt Me 262. This is believed to be the first instance of an Allied aircraft being shot down by a jet fighter. Following this loss, US PR Spitfires were usually provided with a P-51 Mustang escort.

PR XIX
The final and most impressive unarmed PR Spitfire variant to be developed was the PR XIX. Entering service in May 1944, the PR Mk XIX combined the excellent camera fit of the Mk XI

SPITFIRE PR MK XIX
Finished in the standard overall PRU blue finish of late war reconnaissance Spitfires, PM660 was on the strength of 2 Sqn, of 2 Tactical Air Force, based at RAF Fuerstenfeldbruck in south Germany in 1946.

with the Griffon-engine airframe of the Mk XIV. After the first 25 were produced, subsequent aircraft were fitted with the pressurized cabin of the PR Mk X, and the fuel capacity was increased to 1,164L (256 gallons) – some three-and-a-half times that of the original Spitfire. The PR Mk XIX proved even more successful than the preceding PR Mk XI, and by VE Day, the Griffon-powered machine had nearly completely replaced the Mk XI in squadron service.

When production ceased in early 1946, 225 of these exceptional machines had been produced, but the service life of the PR XIX was to extend well into the 1950s.

Mk 21 to the Spiteful

Since 1942, Joe Smith and the Supermarine design team expended a considerable amount of time and effort developing a largely redesigned airframe that best suited the potent Rolls-Royce Griffon. The main change to the basic Spitfire airframe was to the wing, which was subjected to a major redesign for the first time since the prototype was constructed in 1936. The major reason for this change was stiffness. Above certain airspeeds, the wing was flexible enough that when the ailerons were operated, instead of deflecting air upwards or downwards, the pressure of the air caused the ailerons to twist the entire wing, leading to aileron reversal: a condition wherein if the pilot moves the control column to roll to the right, the aircraft rolls to the left.

SPITFIRE MK 21 AND MK 22
A Spitfire Mk 21 (foreground) formates with the Mk 22 prototype, PK312 (centre left). Only the Mk 21 saw action during World War II, and though Mks 22 and 24 were operational in the late 1940s, the jet age had dawned and their service was short.

MK 21 TO THE SPITEFUL

SPITFIRE F MK 21
Only 91 Squadron flew the Mk 21 during the war, operating the type for less than a month and seeing no air-to-air combat. Based at Ludham in Norfolk, 91 Squadron flew armed reconnaissance and anti-shipping sorties during the Mk 21's fleeting operational service.

SPITFIRE F Mk 21
Weight (maximum take-off): 4 167kg (9186lb)
Dimensions: Length 10.03m (32ft 11in), Wing-span 11.25m (36ft 11in), Height 4.11m (13ft 6in)
Powerplant: one 1497kW (2035hp) Rolls-Royce Griffon 61 liquid cooled V-12 piston engine
Maximum speed: 718km/h (446mph)
Range: 933km (580 miles)
Ceiling: 13,045m (42,800ft)
Crew: 1
Armament: Four 20mm (0.79in) Hispano cannon in wings

The theoretical speed at which aileron reversal would occur with the original wing was calculated to be 933.4km/h (580mph) – comfortably in excess of the maximum capable of being attained by early variants, even in a dive. However, with the maximum speed possible in a dive edging ever closer to this figure, the need for a less flexible wing was becoming urgent. The new wing was built with strengthened spar booms and a thicker gauge aluminium skinning that resulted in a 47 per cent increase in stiffness. This raised the theoretical aileron reversal speed to a colossal 1,3272km/h (825mph), safely beyond anything achievable by any Spitfire mark. The ailerons themselves were also extended outwards by 20.32cm (8in), resulting in a straighter trailing edge and requiring a change to the wingtip shape. This represented the first visual change to the Spitfire wing on any production model.

Mk 21

As Mark numbers climbed to ever higher figures, a switch to Arabic numerals was eventually made for the sake of simplicity. The second Spitfire Mk 21 prototype was the first to be fitted with the new wing, and the first prototype, which had started life as the second prototype Mk IV/XX but was then modified to Mk 21 standard. Other changes saw the undercarriage legs relocated further outboard to increase the track by nearly 20.32cm (8in), while the undercarriage was lengthened to allow a larger diameter propeller to be fitted. This was a five-bladed Rotol propeller of 3.35m (11ft) diameter, as opposed to the 3.18m (10ft 5in) diameter propeller of the Mk XIV and XVIII. Half-wheel doors were fitted to the wings to fully enclose the wheels when retracted. Four 20mm (0.79in) cannon were fitted as standard, and the normal engine fitted to the Mk 21 was the Griffon 61 rated at 2035hp (though some aircraft received the Griffon 64 modified to deliver 2375hp). A few examples were eventually flown with the Griffon 85 and a Rotol six-bladed contra-prop.

The Mk 21 prototype flew for the first time in July 1943, and production contracts for 3000 aircraft were placed with the CBAF. A further 400 were to be built by Supermarinehe first production Mk 21 was delivered

MK 21 TO THE SPITEFUL

in January 1944. The Mk 21 was 16.1–19.3km/h (10–12 mph) faster than the Mk XIV at all altitudes, despite its higher weight, but the programme was thrown into jeopardy by a damning report from the AFDU which criticized the lateral handling of the aircraft and its critical trimming characteristics. Shockingly, the report concluded: "In its present state it is not likely to prove a satisfactory fighter. No further attempts should be made to perpetuate the Spitfire family."

This statement would prove premature and led Supermarine test pilot Jeffrey Quill to later state "their first report over-did it a bit in my opinion". An intensive effort followed at Supermarine to alleviate the aircraft's handling problems, which were largely solved by changing the trim tab gearing and other minor control modifications. The changes were successful enough that the Spitfire Mk 21's handling qualities were declared to be acceptable by the AFDU when they re-evaluated the aircraft in March 1945, declaring, in typically dry terms: "It is considered that the modifications to the Spitfire 21 make it a

SPITFIRE F MK 22
In overall RAF 'high speed silver' finish, PK433 was on the strength of 603 (City of Edinburgh) Squadron of the Royal Auxiliary Air Force, as denoted by the coloured band either side of the roundel. Based at RAF Turnhouse in Scotland, 603 Squadron flew the Mk 22 from 1947 until 1951.

satisfactory combat aircraft for the average pilot". The first unit to receive Mk 21s was 91 Squadron, which was one of the two initial squadrons to operate the first Griffon-powered Spitfire Mk XIIs. Despite only just making it into service before the end of hostilities in Europe – 91 Squadron being declared operational a shade under a month before VE Day – the Mk 21 can best be considered the first of the postwar Spitfires, though few were to be built. Despite large orders for the type, the end of the conflict saw most of the production contracts for Mk 21s being cancelled. Ultimately, only 120 were produced, all by the CBAF.

F Mk 22
Joining the F Mk 21 on the production line in the last few

SPITFIRE F Mk 22
Weight (maximum take-off): 5148kg (11,350lb)
Dimensions: Length 10.03m (32ft 11in), Wing-span 11.25m (36ft 11in), Height 4.11m (13ft 6in)
Powerplant: One 1771kW (2120hp) Rolls-Royce Griffon 85 liquid cooled V-12 piston engine
Maximum speed: 724km/h (450mph)
Range: 933km (580 miles)
Ceiling: 13,868m (45,500ft)
Crew: 1
Armament: Four 20mm (0.79in) Hispano cannon in wings; up to 460kg (1000lb) bomb load

weeks of the conflict was the F Mk 22, which differed most obviously from the Mk 21 in its adoption of the cut-down rear fuselage and teardrop canopy already developed for the Mk XIV. Under the skin, a major change was the replacement of the 12-volt electrical system, as used by all previous Spitfires, with a 24-volt system. This variant was produced in slightly greater numbers than its immediate

MK 21 TO THE SPITEFUL

NEW WING
Here, the new wing designed for the Spitfire Mk 21/22/24 can be clearly seen, in this case fitted to a Mk 22. The longer ailerons, reaching nearly to the wingtip, are also evident.

forbear, with 287 eventually being built, most by the CBAF. However, the final 27 were built by Supermarine. Rounding up the Spitfire's production life, the final 54 airframes were built to F Mk 24 standard. The Mk 24 differed little from the preceding Mk 22, again featuring the cut-down rear fuselage and clear vision canopy, but it did possess two 150L (33 gallon) fuel tanks in the rear fuselage and was fitted with four short barrel Hispano Mk V cannon with an electrical firing system. Both the Mk 24 and later production Mk 22s were equipped with the enlarged tail surfaces developed for the final offshoot of the Spitfire story: the Supermarine Spiteful.

Supermarine Spiteful
Development of the Spiteful had begun in 1942 with Supermarine's decision to develop a new wing for the aircraft that would improve rate of roll and overall performance. Two avenues of development were pursued in the form of two different wing designs.

Firstly, the so-called 'minimum modification' wing attempted to achieve improvements whilst retaining the Spitfire's planform and introducing only a minor change in aerofoil profile. This wing was tested on a Spitfire Mk VIII but, although offering some benefit to maximum speed, the improvement was deemed too modest to proceed with.

By contrast, the second wing design was completely new and was intended to achieve 'laminar flow', whereby airflow over the wing is prevented from breaking away from its surface, thus eliminating the turbulent flow of air produced by a standard wing: a significant contributor of aerodynamic drag. To achieve this, Supermarine designed a thinner wing with maximum thickness and placed further back than previously. Such a thin wing could not easily accommodate the mainwheels when retracted, so an inward retracting undercarriage was designed, the main wheels residing in the wing root, the thickest part of the wing, when retracted. This design also eliminated one of the persistent issues still dogging the Spitfire: its unusually narrow track. The new wing also eliminated the Spitfire's distinctive elliptical wing plan with the aim of easing production. Radiators were fitted under the wings, as before, but in a bid to reduce drag, they were of a much shallower and wider design.

In 1943, Air Ministry specification F.1/43 was drawn up to cover Supermarine's work on the new wing and proposed that it be fitted to a Merlin-powered Mk VIII fuselage or Griffon-powered Mk 21. However, the first prototype utilized another fuselage altogether. The name 'Victor Mk II' was provisionally assigned to the project (Victor Mk I being the name originally proposed for the Spitfire Mk 21), but by the time the prototype appeared in June 1944, the definitive name of 'Spiteful' had been adopted.

First flight

The prototype was in fact a Spitfire Mk XIV fitted with the new wing and flew for the first time on 30 June 1944. Unfortunately, the Spiteful programme was subject to much delay, initially due to the loss of the first prototype in a fatal crash only three months after its first flight followed by a wait of another three months before the second prototype was completed. The new prototype was fitted with the definitive Spiteful fuselage, featuring a slightly raised cockpit to improve view over the nose, a large teardrop canopy and cut-down rear fuselage. This aircraft was built to the production standard as then defined and was designated the Spiteful F Mk XIV to signal the close similarities between the new fighter and the equivalent mark of Spitfire. Plans for production were begun, with some existing Spitfire contracts being altered to cover Spitefuls instead.

Spiteful F Mk XIV

Sadly, and in stark contrast to the wing it replaced, the new wing had problematic handling characteristics, with a vicious stall, yet exhibited more problematic compressibility effects at high speed than the Spitfire, in the form of localized airflow exceeding

SUPERMARINE SPITEFUL F MK XIV
This is RB515, the first production aircraft, photographed during flight trials. Although its Spitfire ancestry is still apparent, the distinctive straight-edged wing and large tail surfaces of the Spiteful are obvious.

the speed of sound, causing drag. Valuable time was lost in modifying the prototype to deliver an acceptable standard of handling. An enlarged tail unit was developed, increasing the area of the horizontal surfaces by 27 per cent and that of the fin and rudder by 28 per cent, and proving successful enough that it was fitted to the final Spitfires produced.

The first production Spiteful F Mk XIV flew on 2 April 1945, powered by a Rolls-Royce Griffon 69, with plans in motion for a version with a Griffon 89 or 90 and six-bladed contra-prop to be designated the Mk XV. None of the latter would be produced, however, as with the war coming to a close, and the immense performance potential of the first jets becoming apparent, only 17 Spiteful Mk XIVs would be built from the original order for 800. None of these 17 would ever be issued to a service unit.

Despite this, a few carried on flying for development work, and one example fitted with a Griffon 101 engine, curved windscreen and five-bladed propeller achieved 795km/h (494mph) at 8,473m (27,800ft), believed to be the

SPITFIRE MK IX
The solitary Soviet two-seat Spitfire conversion was carried out by 1 Aircraft Depot, Leningrad in 1945. The Soviet Union regarded the Mk IX Spitfires they received as invaluable high-altitude combat aircraft.

highest speed ever attained by a British piston-engined aircraft in level flight. Development of a naval Spiteful variant would continue, however, as the 'Seafang', as detailed in the next chapter.

Training role
The postwar era also saw the Spitfire adapted to fulfil its last new military role as a dedicated training aircraft. Various Spitfire marks, of course, had been flown as advanced trainers during the war, with some even serving as deck landing training aircraft for the Royal Navy, but a two-seat version with provision for instructor and pupil was limited to a few one-off adaptations, generally at unit level: a Mk Vc was converted as a two-seater by 261 Squadron, for example, and a single Mk IX was similarly altered by the Soviets for training use. The production of factory-converted trainers took

SPITFIRE Mk IX
Weight (maximum take-off): 4309kg (9500lb)
Dimensions: Length 9.47m (31ft 1in), Wingspan 11.23m (36ft 10in), Height 3.86m (12ft 8in)
Powerplant: One 1151kw (1565hp) Rolls-Royce Merlin 61 liquid cooled V-12 piston engine
Maximum speed: 657km/h (408mph)
Range: 698km (434 miles)
Ceiling: 13,106m (43000ft)
Crew: 2
Armament: two 20mm (0.79in) Hispano cannon and four 7.7mm (0.303in) Browning machine guns, or four 20mm Hispano cannon in wings; up to 460kg (1000lb) bomb load

place after 1945, with the first, a converted Mk VIII flying in 1946, serving as a demonstrator for potential customers. Ten two-seat Mk IXs were subsequently ordered by India in 1948, these being converted from standard Mk IXs by Vickers-Supermarine and designated T Mk IXs. A further

MK 21 TO THE SPITEFUL

six T Mk IXs were then produced for the Irish Air Corps in 1951, which was then operating eleven 'denavalised' Seafire Mk IIICs as its sole fighter type. This marked the end of 'new' Merlin-engine Spitfire developments, and a two-seat Griffon Spitfire was never developed. The end of Merlin-powered Spitfire production did not, however, correspond to the end of Merlin-powered Spitfires in service use.

SPITFIRE HF Mk IXC
Weight (maximum take-off): 4309kg (9500lb)
Dimensions: Length 9.47m (31ft 1in), Wingspan 11.23m (36ft 10in), Height 3.86m (12ft 8in)
Powerplant: One 1162kw (1580hp) Rolls-Royce Merlin 66 liquid cooled V-12 piston engine
Maximum speed: 650km/h (404mph)
Range: 698km (434 miles)
Ceiling: 12,954m (42500ft)
Crew: 1
Armament: Two 20mm (0.79in) Hispano cannon and four 7.7mm (0.303in) Browning machine guns, or four 20mm Hispano cannon in wings; up to 460kg (1000lb) bomb load

Soviet defence role

The Mk IX in particular saw widespread postwar service for an array of nations. The Soviet Union, for example, received 1188 Mk IXs from February 1944 through Lend-Lease channels during the war. Impressed with the Mk IX's altitude performance, and cognisant of the Spitfire's relatively poor rough-field capability, the Soviets utilized their Mk IXs for the defence of high-priority targets and cities. As such, due to the paucity of German aircraft penetrating deep into Soviet airspace by this stage of the war, the Mk IXs saw little use, with a single victory in March 1945 when two Mk IXs shot down a Ju 88 over Leningrad at a height unattainable by any other Soviet fighter. After the war, the USSR kept the Spitfire IX in service due to its unrivalled altitude performance, some remaining operational until 1951, by which time the formidable MiG-15 was in service.

SPITFIRE HF MK IXC
Expatriate Danes raised sufficient funds during the war to supply three presentation Spitfires in RAF service, and all were flown by Danish pilots. After the war the Royal Danish Air Force operated 38 HF Mk IXs from 1947 to 1955.

European service

France was another major user of Spitfires. Over 500 Merlin-powered variants were utilized by the Armée de l'Air for several years following VE Day, including in combat on ground attack missions during the initial stages of the Indochina War. Other major European users included the Netherlands, Czechoslovakia, Turkey, Belgium and Greece, the Hellenic Air Force utilizing their Mk IXs in action during the Greek Civil War from 1946 to 1949. Spitfire Mk IXs also proved popular with Italian pilots. Italy flew 145 Spitfire Mk IXs until the early 1950s, and later these Spitfires found their way to Israel, supplementing the ex-Czech aircraft already in service there, and Burma.

Middle East wars

In the Middle East, Spitfire Mk IXs were operated by Egypt, and in a curious twist, Israel, who controversially obtained a fleet

111

of Mk IXs second-hand from Czechoslovakia. During the 1948–1949 Arab–Israeli war, a messy three-way encounter in which some (neutral) RAF Spitfire Mk. XVIIIs were attacked on the ground by Egyptian Spitfires IXs (who had misidentified them as Israeli Spitfires) proved to be the final air-to-air combat involving Spitfires. A later strike by five Egyptian Spitfires resulted in all five attackers being destroyed – three by ground fire and two by British Spitfires (the last of which remains the most recent victory in air combat by an RAF pilot in an RAF aircraft).

Some days later, Israeli Spitfire Mk IXs mistook British Spitfires for Egyptian Spitfires and shot down two. The very final air-to-air Spitfire victory anywhere was scored by American pilot Bill Schroeder in an Israeli Spitfire LF Mk IX on 7 January 1949, when he shot down an RAF Hawker Tempest (again, apparently, mistaking it for an Egyptian aircraft).

Around the same time, the Spitfire Mk IXs of the Indian Air Force were in action against ground forces in Kashmir, India, and would continue to be operated until at least 1957. The final combat use of the Spitfire anywhere in the world occurred in Burmese hands: 30 ex-Czech Mk IXs, 20 Seafire Mk XVs and three Mk XVIIIs were used in counter-insurgency ground-attack missions against Communist separatists in the north of the country until 1954.

Late RAF usage

Postwar RAF usage of the Spitfire was concentrated on Griffon-engine variants, though Mk IX and Mk XVIs were used as advanced trainers for several years after the war. The final combat usage of the Spitfire in British hands occurred during the so-called 'Malayan Emergency' of 1948–1960. Spitfires saw action against Communist ground targets from July 1948 until the last strike was flown by Spitfire Mk XVIIIs on 1 January 1951. Reconnaissance Spitfires persisted on operations for somewhat longer, the final operational sortie by a PR Mk XIX over Malaya taking place on 1 April 1954 from RAF Seletar in Singapore. This aircraft had 'The Last' painted on its nose to commemorate the fact.

ISRAELI SPITFIRES
The first batch of Israeli Spitfires, which included both 03 and 15 depicted below, were ferried in 'Operation Velvetta' from Czechoslovakia to Nikšić in Yugoslavia and then to Israel. Ex-Luftwaffe external fuel tanks were utilized, adapted as underwing tanks.

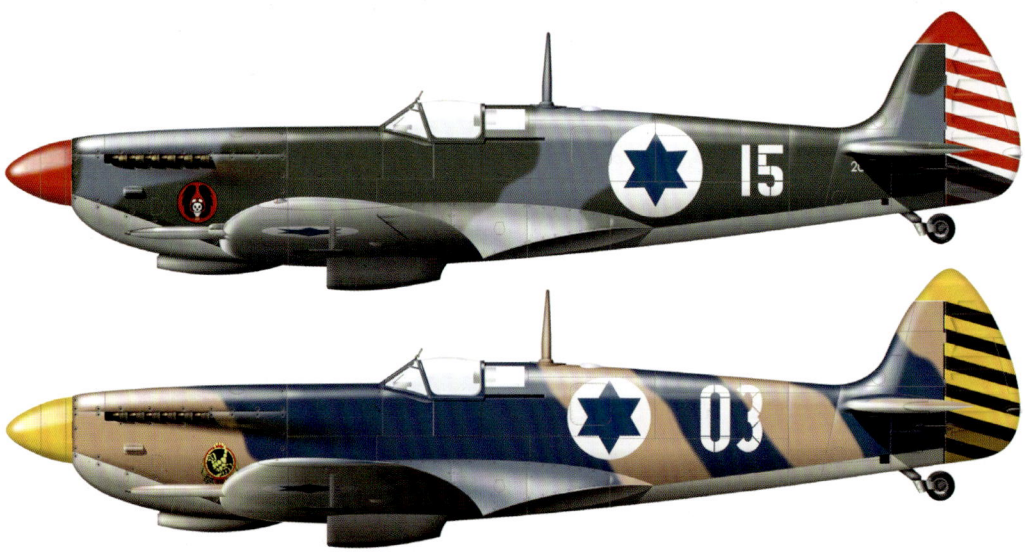

RESTORED MODELS

Since its retirement from air forces around the world, dozens of Spitfires of various marks have been restored to flight, notably with the RAF's own Battle of Britain Memorial Flight. The world's population of airworthy Spitfires at the time of publication is around 50, with this figure having been maintained for several years now as new aircraft are being restored while others are withdrawn either temporarily or permanently. There is no sign that the desire to fly Spitfires is waning – in 2015, a post-service record of 17 Spitfires flew in formation at the Imperial War Museum's Duxford airfield in Cambridgeshire. More recently, in August 2019, a highly polished Mk IX set off from Goodwood Aerodrome in the UK on a circumnavigation flight around the world. The 'Silver Spitfire' covered over 43,452km (27,000 miles) on its journey, which involved 15 major stop-offs, before the 76-year-old aircraft arrived back at Goodwood some four months after it departed. There is little doubt that the unique shape of the Spitfire will continue to grace the skies for decades to come.

BATTLE OF BRITAIN MEMORIAL FLIGHT
One of six Spitfires operated by the RAF's Battle of Britain Memorial Flight, PS915 is one of two airworthy PR XIXs on the airshow circuit, the other is owned by Rolls-Royce. Airworthy Merlin Spitfires hugely outnumber Griffon-powered examples.

The last flight by an RAF Spitfire – also the last known flight of a piston-engined fighter derivative in regular RAF service – was by a PR Mk XIX of the Temperature and Humidity Flight at RAF Woodvale, on 9 June 1957. However, in an odd coda, in 1962, Air Marshal Sir John Nicholls flew a Battle of Britain Memorial Flight Spitfire PR Mk 19 against a supersonic English Electric Lightning F 3 in mock combat at RAF Binbrook. At the time, British forces were considering possible intervention against Indonesia over Malaya, and Nicholls had decided to develop tactics to fight Indonesian P-51 Mustangs, a fighter with similar performance to the PR Mk 19.

The Seafire

In the late 1930s, the Spitfire was arguably the finest fighter aircraft in the world, and the Royal Navy was keen to take the aircraft to sea. As early as 1938, the Admiralty had approached Fairey about the possibility of building navalized Spitfires under licence. However, it was during the following year that the first definitive steps towards the production of a carrier version of the aircraft took place, when Vickers-Supermarine began working with the Admiralty's Advisory Committee on Aircraft to design a folding wing system and arrestor hook for the Spitfire.

FLAT-TOP FIRE
The experience of operating the Spitfire from a carrier was never an entirely happy one. Here a Seafire XVII of 737 Naval Air Squadron has suffered a barrier strike on HMS *Illustrious* in June 1949, nearly wrenching the starboard wing clean off the fuselage. Fire extinguishers ply over the aircraft as the pilot makes a hasty exit from the cockpit.

THE SEAFIRE

At around the same time the Royal Navy regained direct control of carrier aviation, the Fleet Air Arm having been a branch of the RAF for much of the interwar period and as a direct result, the development of naval aircraft had been somewhat neglected. In response to Air Ministry specification N.8/39, issued in 1939, Supermarine proposed a Spitfire-based design featuring Corsair-style inverted gull wings and powered by either the Rolls-Royce Griffon or Napier Sabre. This was rejected on the basis that view over the nose was considered inadequate for deck landing, so the Fairey Firefly was ordered instead.

Navy need

However, a different avenue for Royal Navy Spitfire operation became available when, the day before Supermarine's N.8/39 proposal was rejected, Admiral Guy Royle held a meeting to discuss 'Future Policy for Fighters' in which it was noted that the Royal Navy was now expected to defend naval bases that had previously been the responsibility of the RAF. As the Navy's existing fighters had never been intended to perform this role and were ill-suited to undertake it, the meeting concluded that it would be 'desirable to reinforce the weapons of the Fleet Air Arm with a number of high-speed single-seater fighters of the most modern types', noting this point in a section entitled 'Hurricanes or Spitfires' – a title that makes it plainly obvious which 'modern types' the Navy had in mind.

After the war had begun, the campaign in Norway starkly demonstrated the limitations of the Royal Navy's then-current fighters: the biplane Gloster Sea Gladiator, the clumsy two-seat Blackburn Skua and its even less effective offshoot, the turret-armed Blackburn Roc. Such limited aircraft led to the hasty adoption of a modified Hawker Hurricane adapted for carrier operations in

SEAFIRE MK IIC
Complete with tropical filter under the nose, this Seafire of 899 Squadron is pictured on *Indomitable*'s forward lift in March 1943. Note the trolleys under the main wheels, allowing the aircraft to be pushed onto or off the lift sideways. The non-folding wing Mk IIC meant it could only fit on the lift facing either to port or starboard, not fore and aft.

THE SEAFIRE

SEAFIRE MK IIC
MB270 was serving aboard HMS *Attacker* during operations off Crete in September 1944. A month later, MB270's career ended when it missed the arrestor wires on landing and was written off following the barrier strike.

Prototype
The prototype Seafire was converted from a standard Spitfire Mk Vb equipped with the large Vokes air filter commonly seen on aircraft serving in the Western Desert. This particular machine was a presentation aircraft paid for by the Dutch East Indies and named 'Bondowoso'. After being fitted with an arrestor hook and catapult spools, Bondowoso was used by Lieutenant Commander HP Bramwell to make the first series of deck landings by a Seafire. Sources differ on exactly when this took place, but it was either during the end of December 1941 or early January 1942. Bramwell made 12 arrested landings and 11 take-offs from HMS *Illustrious*, both with and without utilizing catapult assistance. An initial assessment of the Seafire's deck-landing characteristics rated them satisfactory, though concern was raised about the long nose adversely affecting visibility on approach.

The same aircraft was subjected to handling trials at the A&AEE at Boscombe Down during April 1942 where it was, unsurprisingly, found to possess very similar

SEAFIRE MK IIC
Weight (maximum take-off): 3240kg (7145lb)
Dimensions: Length 9.12m (29ft 11in), Wingspan 11.23m (36ft 10in), Height 3.02m (9ft 10in)
Powerplant: One 1055kW (1415hp) Rolls-Royce Merlin 46 series liquid cooled V-12 piston
Maximum speed: 555km/h (345mph)
Range: 793km (493 miles)
Ceiling: 9755m (32,000ft)
Crew: 1
Armament: two 20mm (0.79in) Hispano cannon and four (0.303in) Browning machine guns in wings; up to 115kg (250lb) bomb load

the form of the Sea Hurricane. The Air Ministry, however, continued to resist calls to navalize the Spitfire, mainly due to cost, pointing out that the expense of adapting 50 Spitfires for carrier operations would cost more than building 50 new Spitfires for the RAF. However, by mid-1941, the Sea Hurricane was on the brink of obsolescence, and a sufficient number of the excellent Grumman F4F Wildcat (known at this time as the Martlet in Royal Navy service) could not be obtained from the US due to the US Navy's own pressing needs. Luckily for the Navy, a visit by Prime Minister Winston Churchill to HMS *Indomitable* in September revealed the sorry state of their fighter component, and the following month, Churchill lent his personal support to Admiralty requests for Spitfires. As a direct result of this intervention, work began apace.

handling qualities to the Mk V Spitfire, despite the somewhat greater weight as a result of the naval equipment. The first serially produced Seafire Mk IBs were also minimal conversions of Mk Vb Spitfires, with the most obvious external change being the fitting of an A-frame arrestor hook under the rear fuselage. The extra weight of this item was balanced

THE SEAFIRE

FLIGHT DECK
These Seafire Mk IICs of 880 Squadron are ranged for take-off aboard HMS *Indomitable* as Fairey Albacore torpedo bombers fly overhead during operations in the Mediterranean in 1943.

by two 12kg (26.5lb) lead weights either side of the engine. Unlike the prototype Bondowoso, catapult spools were not fitted. Internal changes included fitting a naval radio and an airspeed indicator calibrated in knots.

Such was the rush to get the Seafire into service that by the time the formal specification and contract for production of the aircraft had been issued in August 1942, most of the initial 48 Seafires had already been delivered.

Seafire II
Work had been progressing concurrently on the development of the Seafire II: a more thoroughly modified aircraft featuring catapult spools to allow the aircraft to be catapult-launched if necessary. These aircraft were built from scratch rather than being converted from existing airframes, though it was still a fairly basic adaptation of the Spitfire V.

The aircraft was known as the Seafire Mk IIC, the 'C' denoting that the Supermarine cannon-armed 'universal' wing was fitted as developed for the Spitfire Mk V. It is worth noting that the Fleet Air Arm also made use of many 'hooked Spitfires' – standard ex-RAF Spitfires fitted with an arrestor hook but lacking any other naval equipment. These were used solely for training and were never referred to as Seafires.

Combat debut
The first unit to convert to Seafires, 807 Naval Air Squadron (NAS), embarked with their aircraft on HMS *Furious* in August 1942. A measure of the urgency with which the Seafire was required can be gauged by the fact that by the end of September, a further four units had re-equipped with the aircraft. The Seafire's combat debut would have to wait until the Operation Torch landings in the autumn of 1942, during which it provided air cover to the invasion force and scored its initial aerial victories. The first of these victories occurred on 8 November, when a Seafire shot down a Vichy French Martin 167.

On operations, the Seafire – though it had inherited the Spitfire's outstanding handling in the air – was gaining a reputation for fragility on deck. The aircraft possessed a gentle stall; an unfortunate trait when performing carrier landings, as the aircraft often tended to float over the arrestor wires to crash into the barrier. The most successful deck landing aircraft tend to possess a more abrupt stall, allowing them to be stalled onto the deck and stay there. Even if all went well and an arrestor wire was caught by a Seafire, the position of the hook under the fuselage resulted in a sharp nose down pitch as the aircraft decelerated; this combined with the Spitfire's minimal airscrew clearance often caused the propeller blades to 'peck' the deck and render the aircraft unserviceable. Even if a pilot managed to avoid those two pitfalls, the undercarriage, inherited from the Spitfire, was not strong enough to withstand the repeated high stress of deck landings, and was prone to collapse. These issues came to an unfortunate head during Operation Avalanche, the Salerno landings of September 1943, when low windspeed over the deck, poor visibility, and inexperience exacerbated the deck-landing woes of the Seafire.

As a result, 73 aircraft from the original force of 105 were lost or seriously damaged to non-combat causes over the course of a mere three days. Nonetheless, the Seafire's airborne performance was sufficiently good for the decision to be made to continue operations with it whilst attempting to mitigate the worst of its deck-landing issues.

Folding wing

By this time, work to provide the Seafire with a folding wing had been carried out, with Supermarine and the Ministry of Aircraft Production collaborating on the design. The system adopted was unusual amongst British naval fighters in that the wings folded vertically upward, though to allow clearance within the limited height of the standard British carrier hangar, the wingtips folded separately, resulting in a 'Z' form when folded. The first production Mk IIc was rebuilt with folding wings in December 1942. Remarkably, the new folding wing was only 8.6kg (19lb) heavier than the standard wing it replaced. Successful flight trials took place in January 1943, and the aircraft was ordered into production as the Seafire Mk III, with deliveries beginning in April. The Seafire Mk III would ultimately become the most numerous Seafire variant, with 1220 produced: 870 by Westland and a further 350 by Cunliffe-Owen.

Seafire III

Three subtypes were produced: the standard F Mk III and the fighter-reconnaissance FR Mk III, which both featured provision for cameras in the rear fuselage and were powered by the Rolls-Royce Merlin 55, and the L Mk III, which utilized the Merlin 55M powering a four-bladed propeller and was intended for low-altitude operations. Many Seafires, like their land-based cousins, were intended for low-

FOLDING WING
Wing-folding allowed Seafire Mk IIIs to be stored below decks on board the smaller of the RN's carriers. Note that a double fold was necessary, such was the restricted space in carrier hangers.

THE SEAFIRE

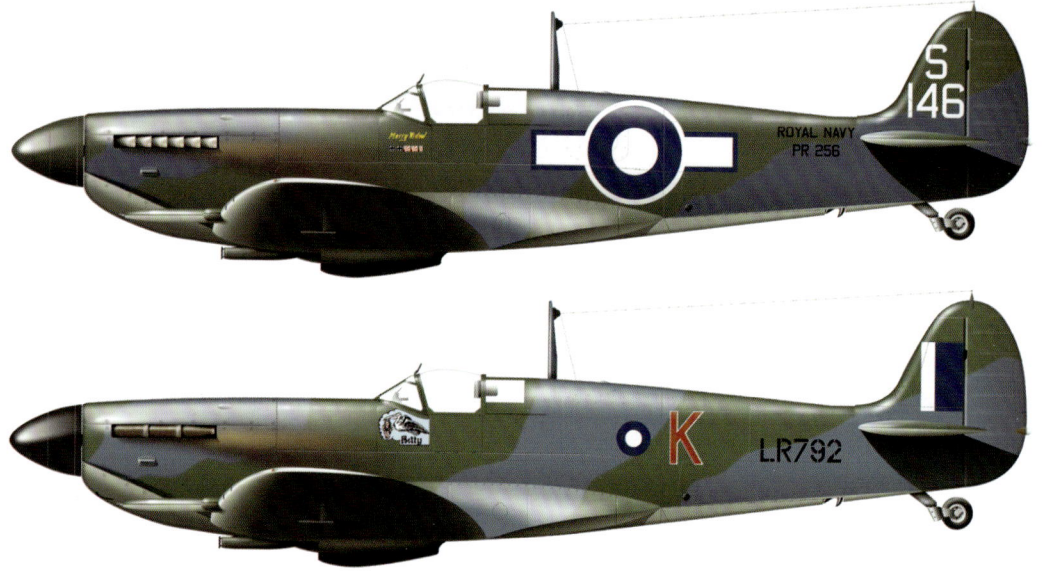

SEAFIRE MK III
Weight (maximum take-off): 3275kg (7220lb)
Dimensions: Length 9.12m (29ft 11in), Wingspan 11.23m (36ft 10in) or 4m (13ft 4in) wings folded, Height 3.02m (9ft 10in)
Powerplant: one 1182kW (1585hp) Rolls-Royce Merlin 55M liquid cooled V-12 piston engine
Maximum speed: 560km/h (348mph)
Range: 748km (493 miles)
Ceiling: 9753m (32,000ft)
Crew: 1
Armament: Two 20mm (0.79in) Hispano cannon and four (0.303in) Browning machine guns in wings; up to 230kg (500lb) bomb load

Top: SEAFIRE L MK III
Serving with 894 Squadron aboard HMS *Indefatigable* off the Sakishima Gunto Islands, PR256 was flown by Sub/Lt Richard Reynolds, the highest scoring Seafire pilot and used by him to shoot down two A6M Zeros on 31 March 1945.

altitude operations and fitted with clipped wings. Later production examples also featured Hispano V short-barrelled lightweight 20mm (0.79in) cannon. By now, deck-landing issues had been much improved, though the Seafire was still more accident-prone than the US-built fighters.

Seafire IIIs undertook Combat Air Patrol duty over the carriers during Operation Tungsten – the Fleet Air Arm's attempt to sink the German battleship *Tirpitz* at anchor in a Norwegian fjord – and performed dive-bombing

Lower: SEAFIRE MK III (HYBRID)
Only 30 or so of these unusual aircraft were built, featuring the Merlin 55 and four-bladed propeller of the Seafire Mk III but fitted with the non-folding wing and three-pipe exhaust of the Mk IIC. This example served aboard the escort carrier HMS Battler in the Indian Ocean during June 1944.

operations in the Aegean, utilizing American 226.8kg (500lb) bombs on the centreline carrier. Most of the Mk III's service, however, took place with the British Pacific Fleet. A Seafire Mk II had earlier been tested against a captured Mitsubishi A6M5 Zero in trials conducted by the US Navy at Patuxent River Naval Air Station. These trials demonstrated that the Seafire possessed a considerable performance advantage over the Japanese fighter at low altitude and that the Mk III was superior still.

Pacific combat

In the Pacific, once again, the Seafire's deck-landing qualities came up short compared to its American contemporaries, but the Seafire was used for CAP duties protecting the carrier group, while the longer-ranged, ordnance-carrying American fighters escorted bombing aircraft and attacked ground targets.

Towards the end of the war, the combat effectiveness of the Seafire III was improved by the adaptation at unit level of 409.15L (90 gallon) Kittyhawk drop tanks to extend the aircraft's somewhat minimal range closer to that of the Hellcats and Corsairs. Although Seafires did accompany the US-built fighters on fighter sweeps over Japanese-held territory on occasion, they were mainly held back for fleet defence. Its outstanding combat performance, particularly in rate of climb, saw the Seafire III in high demand to combat the increasing number of Kamikaze attackers, a mission that came to a head in operations of Okinawa. Despite ever-dwindling numbers of Japanese aircraft, the Seafire acquitted itself well on the few occasions when air-to-air combat took place, and the very last confirmed Allied air-to-air victory of the war fell to the guns of Sub Lieutenant JG Murphy's Seafire L Mk III operating from HMS *Indefatigable* on the morning of 15 August 1945.

SEAFIRE MK XV

The troublesome Mk XV was a menace on carrier decks but enjoyed a fairly lengthy land-based service life. UB-403 is one of the 20 examples supplied to Burma in 1952, it was withdrawn from service in either 1957 or 1958.

Seafire Mk XV

Meanwhile, a concerted effort was underway to get a Rolls-Royce Griffon-powered Seafire onto the Royal Navy's carriers, Supermarine having proposed a Griffon Seafire as early as January 1943. This development was entirely logical, the Griffon having originally been developed at the behest of the Navy, and specification N.4/43 was issued in August 1943 covering the re-engined aircraft and calling for three prototypes and three pre-production aircraft to be produced. Designated the Seafire Mk XV, the new machine utilized the forward fuselage and four-bladed propeller of the Spitfire Mk XII, the central fuselage and wings of the Seafire Mk III and the rear fuselage and tail surfaces of the Spitfire Mk VIII.

The Griffon-powered Seafire Mk XV just failed to enter service before VJ Day. This may have been something of a blessing, as the Seafire XV developed an appalling reputation, primarily due to its potentially uncontrollable swing on take-off – a result of the switch to the Griffon engine, with its opposite propeller rotation to

FOREIGN SERVICE

French Seafire Mk XV usage would prove brief: having taken delivery of 48 Seafire L Mk IIIs in 1946, the French Navy flew them from the carrier Arromanches. These aircraft saw operational service with the Aviation Navale during the First Indochina War, flying ground-attack missions against Viet Minh targets both from Arromanches and shore bases. Withdrawn from combat operations in January 1949, the aircraft were replaced by Seafire Mk XVs once the carrier had returned to European waters, but the unpleasant characteristics of the Griffon-powered aircraft saw it hastily replaced in French service by the Grumman F6F Hellcat in 1950.

The Burmese Air Force also operated 20 Seafire Mk XVs but flew them exclusively from land bases, supplementing Spitfire Mk IXs on counter-insurgency missions and becoming the last Seafires in front-line service anywhere when the last was withdrawn in 1957, the Irish Air Corps having withdrawn the last of their similarly land-based Seafire L Mk IIIs in 1955.

In total, 390 Seafire XVs were built by Cunliffe-Owen and Westland from late 1944, following the construction of six prototypes by Supermarine.

SEAFIRE L MK III

The Irish An tAerchór (Air Corps) obtained 12 Seafire L Mk IIIs during 1947. These aircraft were stripped of their naval equipment and the wings pinned in the unfolded position, resulting in an aircraft basically the same as the Spitfire Mk Vc.

the Merlin. When a Seafire Mk XV took off, the torque from the propeller tended to pull the aircraft inexorably to starboard, towards the 'island' structure of the carrier, with obviously problematic results. The Mk XV also had inadequate undercarriage that was prone to collapse (it utilized the same landing gear as the Seafire III but weighed considerably more).

Pilots experienced handling problems not just when deck landing but (for the first time) during normal flight, resulting in numerous accidents and groundings. Despite this, the Mk XV served with the Royal Navy, the Royal Canadian Navy and the French Navy.

In British service, the aircraft's various problems meant that its carrier career was short, though it served until 1951 with shore-based second-line units, where the aircraft's problematic qualities for carrier operations were of little consequence. By contrast, the Canadians persisted with operating the Mk XV at sea until replacing it with the superlative Hawker Sea Fury in 1948.

Seafire Mk XVII

The unfortunate experience of the Mk XV led to the rapid development of the much-improved Mk XVII (or F Mk 17, as it was redesignated in 1945 when Arabic numerals replaced Roman numbers). This variant adopted the teardrop canopy and cut-down rear fuselage and combined it with the shorter nose required by the 'short-block' Rolls-Royce Griffon

VI resulting in a highly distinctive profile. Less visually obvious but of great importance, the strengthened undercarriage now featured longer oleos with a slower rebound rate, considerably lessening the tendency of the propeller blades to 'peck' the deck on arrested landings.

One improvement that occurred during the Mk XV's production life was the switch from an A-frame arrestor hook mounted under the fuselage to a 'sting' type fitted at the extreme rear of the aircraft under the rudder. This prevented the aircraft from pitching nose-down so severely when landing on deck. The wings were stronger, extra fuel was carried in a 150L (33 gallon) tank in the rear fuselage and a heavier underwing load could be carried than any previous Seafire variant.

Despite these improvements, the F Mk 17 was only used by two squadrons at sea, 807 NAS taking the F Mk 17 on its initial cruise aboard HMS *Vengeance* in May 1947 but re-equipping on its return. 800 NAS was the only unit to operate the F Mk 17 for any length of time at aboard a carrier, operating from HMS *Triumph* in the Mediterranean between February 1947 and March 1949.

F Mk 17s persisted for longer in shore-based units, however, remaining in frontline service with 1833 squadron of the Royal Naval Volunteer Reserve (RNVR) until 1953. With this, they became the Seafire variant with the longest frontline career, having first entered service in November 1945.

F Mk 45

The development of the Spitfire Mk 21 resulted in an equivalent naval version being produced, the F Mk 45, which utilized the Griffon 60 series and featured the sting-type arrestor hook. Considered an interim type, the F Mk 45 was fitted with a standard Spitfire Mk 21 wing without wing folding. Only 50 were built at the Castle Bromwich Aircraft Factory, entering service with 778 NAS in November 1946. A few were modified to carry two F24 cameras in the rear fuselage, becoming FR Mk 45s. Following the Mk 45, the F Mk 46 was a navalized Spitfire Mk 22, and as such featured the cut-down rear fuselage and teardrop canopy. Once again, the

NO. 800 SQUADRON
Seafire Mk 17s of No. 800 Sqn (and Fairey Firefly Mk Is of No. 827 Sqn) are run up prior to operations from HMS *Triumph* in the Mediterranean in about 1948.

THE SEAFIRE

SEAFIRE F MK 47

VP461 was flown by 800 NAS from April 1948 and flew missions over Malaya and Korea. It is shown as it appeared in August 1950, when it was operating from HMS *Triumph*, armed with underwing rockets and fitted with RATO gear above the wing.

SEAFIRE F Mk 47
Weight (maximum take-off): 5683kg (12,530lb)
Dimensions: Length 10.46m (34ft 4in), Wingspan 11.25m (36ft 11in), Height 3.88m (12ft 9in)
Powerplant: One 1752kW (2350hp) Rolls-Royce Griffon 88-series liquid cooled V-12 piston engine
Maximum speed: 727km/h (452mph)
Range: 652km (405 miles)
Ceiling: 13,135m (43,100ft)
Crew: 1
Armament: Four 20mm (0.79in) Hispano cannon in wings; up to eight 27 kg (60lb) RP3 rockets or 680kg (1500lb) bomb load under wings

wing was non-folding, and the aircraft featured the rear fuselage fuel tank as used by the F Mk 17. The wing was also fitted with plumbing, allowing for the carriage of a 102.3L (22.5 gallon) external fuel tank under each wing in addition to a 227.3L (50 gallon) tank under the fuselage. Nearly all the Seafire Mk 22s were fitted with the larger tail surfaces developed for the Spiteful – a modification that finally eliminated the dangerous swing to starboard of Griffon-engine Seafires. Torque was then totally removed by the 1947 decision to fit all Seafire Mk 46s with contra-rotating airscrews powered by 80 series Griffons. Only 24 were built, however, of 200 originally ordered, all by Supermarine.

F Mk 47

The definitive Seafire, the F Mk 47, entered service in 1948. Fitted with Rotol contra-props from the start, all but the first 14 were powered by the fuel-injected Griffon 88 rated at 2350hp. With external stores, the aircraft could now carry 1123L (247 gallons) of fuel as compared to the 386.4L (85 gallon) capacity of early Seafires. The wing could be hydraulically folded, though the separately folding wingtips were now dispensed with, and the aircraft featured a long air intake, extending to an inlet just below the spinner.

Like the F Mk 46 before it, several aircraft were fitted with F24 cameras in the rear fuselage to become the FR Mk 47.

Tested by the Naval Air Fighting Development Unit in late 1947, the increased weight and power of the F Mk 47 had inevitably caused its handling qualities to degrade, the report noting that due to the oversensitivity of the aircraft in pitch, 'the pilot's attention is constantly required in keeping the aircraft in an accurate climbing attitude'. Moreover, the rudder was overly sensitive, and the aircraft was described as 'uncomfortable to fly' in turbulence.

Despite these concerns, the NAFDU also stated that it was "the best high altitude fighter of all the piston engined aircraft now in service". Although only 90 would be built, the Seafire Mk 47 would see considerable action, initially over Malaya during Operation Firedog in 1949 and later in the Korean War. During the latter, Seafire 47s of 800 NAS would take part in the first air strikes of the war when nine Seafires from HMS *Triumph* attacked the airfield at Haeju with rockets. *Triumph* carried only 12 Seafires,

SEAFANG

A final Seafire derivative deserves to be mentioned: the remarkably named 'Seafang'. The Seafang was a naval development of the Supermarine Spiteful and featured that aircraft's laminar flow wing. The abrupt stall of the Spiteful wing could be seen as an advantage for a carrier aircraft, as it eliminated the Seafire's problematic tendency to 'float' on landing and miss arrestor wires. Equipped with a sting-type arrestor hook, the Seafang flew for the first time in early 1946, though initially in interim Mk 31 form with non-folding wings, of which 150 examples were ordered but only nine would be completed.

This was followed in June 1946 by the first flight of the definitive navalized Seafang Mk 32 with power-folding wings and a contra-rotating propeller. The end of the war extinguished the need for large numbers of naval fighters, but doubts about the practicality of operating jet fighters from aircraft carriers saw the Royal Navy persist with the acquisition of new piston-engined fighters.

Unfortunately for the Seafang, it was perceived not to offer enough of an improvement over the Seafire F Mk 47 to disrupt production, and the rival Hawker Sea Fury, which possessed better low-speed handling, was the preferred choice for the Navy's last piston-engined fighter. The Seafang flew on as a development aircraft for the Supermarine Attacker, the Royal Navy's first jet fighter to enter service, and in this form, a part, at least, of the Seafang lived on, as the Attacker used the same laminar flow wing.

SUPERMARINE SEAFANG F MK 31
The raised cockpit and wide-track undercarriage of the Spiteful were improvements over the Seafire for deck-landing and the Seafang was incredibly fast but the Sea Fury was the better all-round naval fighter.

but the carrier HMS *Unicorn* brought 14 more into the theatre, along with repair facilities, allowing 800 NAS to maintain operations despite inevitable combat attrition. During their time in Korean waters, Seafires flew fighter patrols, reconnaissance, gunnery spotting missions and CAP (combat air patrol) duties. By the time *Triumph* was relieved by HMS *Theseus* in September 1950, only one of her Seafires remained serviceable, this representing the end of the Seafire's career at sea.

Index

AB264 (Spitfire Mk VB) **58**
AB502 (Spitfire Mk VB) **51**
Aboukir 57, 58
    Maintenance Units (MUs) 57, **97**
Admiralty 115, 117
    Advisory Committee on Aircraft 115
Aero-Vee dust filter 69
Aeroplane & Armaments Establishment, RAF Martlesham Heath **25**
Africa 52
aileron reversal 104, 106
Air Defence of Great Britain (ADGB) 69
Air Fighting Development Unit (AFDU) 67, 86–8, 107
Air Force Scientific Institute **60**
Aircraft and Armament Experimental Establishment (A&AEE) 36, 90, 117
airscrews 36–7
Alexandria 57
Algeria 53
Allies 6, 59, 61, 70, 88, **101**, 103
Appledram Airbase **66**
AR213 (Spitfire Mk VB) **6**
Arab-Israeli war 111
Argus (aircraft carrier) 53
Armée de l'Air 111
Arnhem 99
Arromanches (aircraft carrier) 122
Attacker (escort carrier) **117**
Australia 8, 76, **56**, 58
Axis forces 6, 59
Azores **61**

Balkans 61
Barking Creek, battle of 34–5
Battler (escort carrier) **120**
Beardmore engine 16
Beaverbrook, Lord 44, 82
beer flights **0**, 70, **70**
Belgium 111
Benson **102**
Berlin 97, 101
'Betty Jane' (Spitfire Mk VIII) **72**
Biard, Henri 20
Blackburn
    Roc 116
    Skua 116
'Bluebird' car and boat 21
'Bondowoso' (Seafire prototype) 117
Borneo **72**
Boscombe Down 117
Boulogne 97
Boulton Paul Defiant 42–3
Bournemouth 15–16
Bramwell, HP Lieutenant Commander 117
Brest Harbour 50, **101**
Bristol Blenheim **41**, 92–3
Bristol Mercury engine 21
Britain, battle of 6, 27–44, **27**, **38**, 41, **42**, 48, 96
    Memorial Flight 113, **113**
British Air Forces of Occupation (BAFO), No. II (AC) Sqn **87**
Broadhurst, Air Vice Marshal Harry **73**
Broussac, Marcel 98
Browning machine guns 38, 52, 57, 68, 85, 99, 28
BS152 (Spitfire F Mk IXC) **66**
BS459 (Spitfire F Mk IXC) **64**
BSA 28
Burma 60, 73–6, **76**, 111, 112, **121**
Burmese Air Force 122

Calshot 21
Cameron, Flying Officer Lorne **66**
Camoutint **93**, 94
Campbell, Malcolm 'Bluebird' 21
Canada 8
Casablanca **66**
Castel Volturo **72**
Castle Bromwich **6**, 32, 43
Castle Bromwich Aircraft Factory (CBAF) 43–4, 51, 67, 69, 76, 106–8, 123
Checketts, Wing Cdr John **56**
Christie, Flying Officer George Patterson 96–8
Churchill, Winston 42–3, 64, 117
Communists 112
Corsica **66**
Cotton, Sidney 93–5, 96–8
'Cottonizing' 94
Crete **117**
Crimea **60**
Cunliffe-Owen 119, 122
Curtiss CR-3 floatplanes 17
Czech units **42**
Czechoslovakia 111, **112**

D-Day **56**, 61, **62**, 69, 70, 84, 88, **101**
Daimler Benz
    DB 601A engine 61
    DB 605A engine 61
'Dambusters' raid **96**, 99
Darwin, Australia **56**, 58
de Havilland 36–7
    Mosquito 88
Desert Air Force **73**
    No. 7 Wing **54**
Dodecanese 79
Donaldson, Squadron Leader Edward 35
Dowding, Air Chief Marshal Hugh 35, 64, 93–4
DP845 (Spitfire Mk IV) **83**, **84**
Dunkirk 35–6, **37**
Dutch East Indies 117
Duxford **38**, 113

Eagle (aircraft carrier) 53
Eastern Front 60
Eastleigh 6, **10**, **25**, **64**, **101**
Edward VIII **28**
Edwardes-Jones, Flt Lt. Humphrey 25, **25**
Egypt 57, 79, **97**, 111–12
Egyptian Air Force 9
ejector exhausts 27, **29**
Electric Lightning F3 113
EN654 (Spitfire PR Mk XI) **90**
EP356 (Spitfire Mk VB) **60**
Eyston, George 21

F8 cameras 98
F24 cameras 86, 94, 95, 97, 98, 99, 123
F52 cameras 101
Fairey 115
    Albacore **118**
    Firefly 116, **123**

Farnborough 29, 80, 82, 83, 90
Finlay, Squadron Leader Donald 43
Focke-Wulf Fw 190 58, 62, 64, 67, 68, 70, 80, 82–3, **83**
    Fw 190A 86
    Fw 190A-3 67
    Fw 190A-4 **66**
Folland Aircraft Ltd 79
France 8, **8**, 39, **42**, 51, 59, 61, 67, 70, **70**, 84, 88, 92, **93**, 94–5, 98, 111, 122, **62**, 88
    battle for 35
Free French **42**, 73
French Navy 122
French units **47**, **62**
'friendly fire' incidents 34–5
Furious (aircraft carrier) 118

G-load 39
Gabszewicz, Group Captain Aleksander **78**
German aircraft 34–5, 42–3, 68
German forces 54, 79
Germany 23, 29, 79, 94–5, **103**
Gibraltar 53
Gibson, Guy **96**
Gleed, Wing Cdr Ian **48**, **51**
Gloster
    Gladiator 21
    Meteor **41**
    Sea Gladiator 116
Gneisenau (battleship) 50
Goodwood 113
Greece 111
Grumman
    F4F Hellcat 122
    F4F Wildcat (Martlet) 117

Hargreaves, Walter 15
Hawker
    Hart 23
    Hurricane 6, 25, 28, 34–41, 43, 60, 79, 116–17
    Sea Fury 122, 125
    Sea Hurricane 117
    Typhoon 70, 78, 80, 82, **83**
Heinkel He 111 **36**
Hellenic Air Force 111
Heston Aircraft Company 94, **94**, 97, 98, 99
Heston Flight 93–6
Hill, AR 61
Hillborn, Lt Robert 103
Hispano cannon 32, **68**, 85, 108, 120
Hives, Ernest 67
Hooker, Stanley 64–7
Hornchurch **33**, 67

Illustrious (aircraft carrier) 115, 117
Imphal, India **75**
Indefatigable (aircraft carrier) **120**, 121
India **75**, 112
Indian Air Force 60, 112
Indian Ocean **120**
Indochina War 111
Indomitable (aircraft carrier) **116**, 117, **118**
Irish Air Corps 110, 122, **122**
Israel 9, **9**, 111, 112, **112**
Israeli Hel Ha-avir 9
Italian Co-Belligerent Air Force 61
Italy 8, **51**, 59, 61, **72**, 73, **76**, 77, 111

Jannello, Guido 15
Japan 79, 88, 121
Japanese aircraft 58
Japanese forces **75**, 76
Johnson, Wing Cdr James 'Johnnie' 64, **67**
Jones, Veronica Owen **12**
Junkers 23
    Ju 52/3m **53**
    Ju 86 57–8
    Ju 86P 70
    Ju 86P-2 57, 58
    Ju 86R-1 57
    Ju 88 35
    Ju 98 'Stuka' 21

K4049 (B.9/32 bomber prototype (Wellington)) **10**
K5054 (Type 300 prototype) 6, **10**, 21, **22**, 24–5, **25**
K5780 (walrus amphibian) **10**
K7556 (pre-production Wellesley bomber) **10**
K9787 (Spitfire Mk I) 32
K9797 (Spitfire Mk I) **35**
Kai Tak, Hong Kong **88**
Kent, Squadron Leader John **41**
Korean War 124–5, **124**
Kos 79

Labuan **72**
Lamblin radiators 20
'laminar flow' 24, 108
land speed records 21, 34
Lane, Brian **38**
Lavochkin fighter 60
Lend-Lease arrangements 73, 111
Leros 79
Lion 450hp engine 15
Lippisch, Alexander 23
Livingstone **71**
Lockheed 12A 93
Longbottom, Flying Officer Maurice 'Shorty' 93, 94
Ludham **106**
Luftwaffe 8, 36, 37, 41, 61, 83, 84
Lympne **86**
Lynch, Squadron Leader JJ 53

Machowiak, Sergeant Marcin **44**
Malan, Adolf 'Sailor' **33**, **36**
Malaya 112, 113, 124, **124**
Malayan Emergency 112
Malta 53–5, **58**
Marshall Mk XII supercharger 69
Martin 167 118
Martindale, Squadron Leader Tony 90–2
Martlesham Heath **28**
MB270 (Seafire Mk IIC) **117**
McCorkle, Colonel Charles M **72**
McLean, Sir Robert 32
McNair, Pilot Officer 'Buck' **58**
Mediterranean 53, **118**, 123, **123**
Messerschmitt 61, 62
    Bf 109 37, 39, 41, **43**, **44**, 68
    Bf 109F 62
    Bf 109G 61, **68**
    'Komet' 23
    ME 109G 86
    Me 209 V1 34
    ME 262 103
MG FF/M cannon 39
MH434 (Spitfire Mk IXB) **6**
Middle East 52, 68, 73, 89, 111–12

INDEX

MiG-15 111
Ministry of Aircraft Production 44, 82, 119
Mitchell, Reginald Joseph 6, 10, 15–17, **18**, 20–4, **21**, 29
Mitsubishi
    A6M5 Zero 120
    A6M Zero **120**
MK959 (Spitfire LF Mk IXC) **62**
Moehne Dam **96**, 99
Monaco 96
Morotai **71**
Morris, William, Lord Nuffield 32, 43, 44
Morris Motors 32
Mount Farm **102**
Murphy, Sub-Lt JG 121
Mustang 8
Mustang Mk III 86

N3071 (Spitfire PR Mk Type 1A) **93–4**
Nakajima Ki-43 **76**
Napier 15
    Dagger engine 23
    Lion engine 16–17, 20
    Sabre engine 116
Naval Air Fighting Development Unit (NAFDU) 124
Netherlands 77, 111
Nettuno Airfield **77**
New Guinea 76
Newbury, Richard **86**
Nicholls, Air Marshal Sir John 113
NM821 (Spitfire FR Mk XIVE) **87**
Normandy **62**, 88
North Africa 59, 73
North Weald **50**
Norway 79, 99, 116, 120

Okinawa 121
Operation Avalanche 119
Operation Chastise 99
Operation Dragoon 73
Operation Firedog 124
Operation Market Garden 99
Operation Overlord 69, **101**
Operation Sunrise 50
Operation Torch 59, 118
Operation Tungsten 120
Operation Velvetta **112**
Orford 32
Orkneys 58, 72

P7666 (Spitfire Mk IIA) ('Observer Corps') **43**
P9374 (Spitfire Mk IA) **37**
P-38 Lightning, F-5 photo-reconnaissance version 101
P-39 Airacobra 60–1
P-47 59
P-51 Mustang 77, 103, 113
P-51B Mustang 73
P-51C Mustang **77**
Pacific arena 58–60, 79, 120–1
Packard Merlin 266 76
Palestine War 1948 **9**
Patuxent River Naval Air Station 120
Payn, Major Harold 'Agony' 29
Pemberton-Billing, Noel 12–13, **12**, 32
Pemberton-Billing Ltd 12–13
    see also Supermarine Aviation Works Ltd.

'Phoney War' period 32–4
photo reconnaissance 34, 90–103
Photographic Reconnaissance Unit (PRU) 34, 93–6, 98, **101**
Pickering, George 25
Pidgeon, Flight Lieutenant DA **75**
PK312 (Spitfire Mk 22 prototype) **104**
PK433 (Spitfire Mk 22) **107**
PL914 (Spitfire PR Mk XI) **102**
Plymouth 70
PM660 (Spitfire PR Mk XIX) **103**
Poland 39
Polish units **44**, **64**
Portugal **61**
PR256 (Seafire L Mk III) **120**
Pressed Steel company 32
Prinz Eugen (cruiser) 50
PRU blue 96, 99, **103**
PS853 (Spitfire PR Mk 19) **92**
PS888 (Spitfire PR Mk XIX) **103**
PS915 (Spitfire PR Mk XIX) **113**

Quill, Jeffrey 25, 32, 72, 82, **90**, 107

R7059 (Spitfire PR Mk IG) **101**
R7059 (Spitfire PR Type G (later PR Mk VII)) **98**
R.A.E. Restrictor 39, 55
RAF Biggin Hill **27**
RAF Binbrook 113
RAF Fowlmere **23**
RAF Furstenfeldbuck **103**
RAF Kenley **66**
RAF Northolt **44**, **64**, 70
RAF Seletar 112
RAF Turnhouse **107**
RAF Woodvale 113
RB515 (Spitfire Spiteful F Mk XIV) **109**
Red Army Air Force 60
Regia Aeronautica, 20[0] Gruppo 61
Reynolds, Sub/Lt Richard **120**
Reynolds Tubes 32
Robb, Air Vice Marshal James M **78**
Rolls-Royce 17, 20, 51, 67
    Buzzard engines 20
    Goshawk engines 21, 22–3
    Griffon engines 9, 17, 50, 80–9, **80**, 101, 103–4, 106–7, 109, 111–12, **113**, 116, 121–2, 124
    60 series 84–5, 123
    Kestrel engines 17, 20, 23
    Merlin engines 6–8, 17, 22–3, 34, 39, 55–7, 60, 72, **77**, 82, 84, 85, 90, **97**, 109–11, **113**, 121
    45-series 50, 55, 56, 99
    45M-series 56
    46-series 55, 56, 57
    50-series 55, 56
    50A-series 56
    50M-series 56
    55-series 55, 56, 119, **120**
    55M-series 56, 119
    56-series 56
    60 series 64–7, 69, 84, 99
    61 67, 69, 73
    63 69, 73
    64 69
    65 69
    66 73, 76
    70 69, 73
    71 69
    C 27

F 27
I 27
II 27, **29**, 43, 99
III 43
R.M.3SM 44
XII 34, 43
XX 44
'R' engine 20–1
Rotol 106, 124
Royal Air Force (RAF) 6, 9, 10, 21, **24**, 25, 29, 32, 37, 73–7, 79, 80, 83, 92, 94, 112, 116–18
    1 RAF Photographic Development Unit 93
    2nd Tactical Air Force (2TAF) 69, 70, 88, **103**
    'Anti-Diver' patrols **86**, 88
    Fighter Command 35, **42**, 58, 64, 69–70, 93
    'leaning into France' 48, 51
    No 131 (Polish) Wing **78**
    postwar Spitfire use 112–13
    Squadrons
        'Eagle' squadrons 59
        No. 19 Sqn **23**, 29, 32, **35**, **38**, 50
        No. 28 Sqn **88**
        No. 41 Sqn **43**, 84, **85**
        No. 43 Sqn **77**
        No. 54 Sqn **56**, 58, 76
        No. 64 Sqn 67
        No. 65 Sqn **31**
        No. 66 Sqn 32
        No. 74 Sqn **33**, 34–5, 78
        No. 91 Sqn **106**, 107
        No. 92 Sqn **37**, **41**, 51, 73
        No. 124 Sqn 58, 70
        No. 131 Sqn 70–2
        No. 145 Sqn 73
        No. 151 Sqn 35
        No. 152 Sqn **76**
        No. 154 Sqn **64**, 72
        No. 165 Sqn **62**
        No. 212 Sqn 95, 96
        No. 222 **50**
        No. 249 Sqn **53**, 54–5, **58**
        No. 302 (Polish) Sqn **62**
        No. 303 (Polish) Sqn **56**
        No. 306 'Torunski' Sqn **44**, **64**
        No. 310 (Czech) Sqn **66**
        No. 312 Sqn **42**
        No. 329 Sqn **62**
        No. 340 Sqn **42**, **47**
        No. 515 Sqn 99
        No. 541 Sqn **101**, **101**
        No. 542 Sqn **101**
        No. 601 Sqn **48**, **86**, 88
        No. 607 Sqn **75**
        No. 610 Sqn 27, **86**, **86**
        No. 616 Sqn 58, 70–2
        No. 680 Sqn **97**
Royal Aircraft Establishment (RAE) 34, 39, 80–2, 95
Royal Aircraft Factory 13
Royal Australian Air Force (RAAF) **56**, 58–9
    1st Tactical Air Force **72**
    80 Fighter Wing **72**
    No. 1 Fighter Wing 76
    Squadrons
        452 Sqn 76
        457 ('Grey Nurse') Sqn **71**
        457 Sqn **72**, 76
Royal Auxiliary Air Force

(RAuxAF), No. 603 Sqn **107**
Royal Canadian Air Force (RCAF) **58**, **68**, 88
    No. 402 Sqn **56**, **66**
    No. 451 Sqn **66**
Royal Canadian Navy 122
Royal Danish Air Force **111**
Royal Naval Volunteer Reserve (RNVR) 123
Royal Navy 35, 110, 115–16, 121–2, 125
    Fleet Air Arm 82, 116, 118, 120
    Pacific Fleet 120
    Squadrons
        737 Naval Air Squadron (NAS) **115**
        800 Naval Air Squadron (NAS) **123**, 123–5, **124**
        807 Naval Air Squadron (NAS) 118
        827 Naval Air Squadron (NAS) **123**
        880 Naval Air Squadron (NAS) 118
        894 Naval Air Squadron (NAS) **120**
        899 Naval Air Squadron (NAS) **116**
Royle, Admiral Guy 116

S1595 (Supermarine S.6B) **20**
St. Eval **101**
Salerno landings 119
Salonika **97**
Savoia S.13 15–16
Scharnhorst (battleship) 50
Schneider, Jacques 15
Schneider Cup 15–16, **16–18**, 20–1, **20**, 23
Schroeder, Bill 112
Scott-Paine, Hubert 13, 15, 16
Seclin, France **93**, 94
Shaibah Airfield, Iraq **59**
Shenstone, Beverly 23, 24
Shetlands 58
Shilling, Beatrice 39
Shoulidce, Sergeant G **68**
Sicily **54**, 59, 73
Siegfried line 94
Singapore 112
Sinthe, Burma **76**
Sly, Flight Lieutenant Edward **72**
Smith, Flying Officer Len **76**
Smith, Joseph 29–31, 64, 104
Sopwith, 'Bat Boat' 12
South Africa 61
South African Air Force (SAAF)
    No. 2 Sqn **54**
    No. 40 Sqn **51**
South East Asia Command **76**
Soviet Union 8, 60–1, **60**, 99, 110–11, **110**
    1 Aircraft Depot **110**
Stainforth, Flt Lt George **20**
Standford-Tuck, Robert **31**
Stokoe, Jack **43**
Suez Canal 57
Summers, Joseph 'Mutt' 6, **24**, 25, **25**
Supermarine Attacker 125
Supermarine Aviation Works Ltd. 10, 12, **13**, 15
    see also Pemberton-Billing Ltd
Supermarine Baby 15
Supermarine Nighthawk 13–15, **13**

INDEX

Supermarine S.4 17–21
Supermarine S.5 17–21, **18**
Supermarine S.6 21
S.6B **20**, 21
Supermarine Sea King 16–17
Supermarine Sea Lion 15, 16
II 16, **17**, 20
III 17
Supermarine Seafang 125, **125**, 110
Supermarine Seafire 8, 9, 115–25
armaments **117, 120, 124**
F Mk 45 123–4
F Mk 46 123–4
F Mk 47 124–5, **124**
folding wing 119, **119**
Mk II 120
Mk IIC **116–18**, 118–19, **120**
Mk III 119–21, **119, 120**
F Mk III 119
FR Mk III 119
L Mk III 119, **120**, 121, 122, **122**
Mk IIIC 110
Mk XV 112, 121–2, **121**
Mk XVII (F Mk 17) 122–4, **123, 115**
prototype 117
Supermarine Spitfire
armaments 28, 32, **33**, **35**, **37–8**, 39, **42–4**, **51–2**, 52, **54**, **56**, **58**, 60, **60**, **64–5**, **71–2**, **76**, 78, **78**, **85–7**, **94**, **100**, **102–3**, **106–7**, 108, **110–12**
body 24
bomb shackles/racks 54–5, 69, 73
C-type ('Universal') wing 52, 68, 76, 85
camouflage 28–9, **29**, **42**, **64**
canopy jettison system 57
'clipped wing' 56
cockpit knock out panels 28
cockpits **45**, 57
coolant systems 34, 50
costliness 24
dust filters 51, 52
'E' wing **67**, 68–9, 76, 85
early production 31–2
elliptical wing 23–4, **23**, **52**
first combat 32–6
flush rivets 23–4
High Speed Spitfire 34, **34**
'hooked Spitfires' 118
'minimum modifcation' wing 108
Mk I 27–44, **31**, **33**, **35**, 39, 44, 51, 79, 93–4, 99, 101, **101**
Mk I PR Type B 95–6, **95**, 97
Mk I PR Type C ('Long Range' 'LR' Spitfire, later PR Mk

III) **95**, 96–8, 99
Mk I PR Type D ('Extra Super Long-Range SPifire') 97, 98
Mk I PR Type D ('Extra Super Long-Range Spitfire', later PR Mk IV) 97, 98
Mk I PR Type E (later PR Mk V) 97, 98, 99
Mk I PR Type F (later PR Mk VI) 97, 98
Mk IA **23**, **37**, 50, 67, **93**
Mk IB 32, **38**, 39, 117
Mk II **42**, 43–4, 51, 101
Mk IIA **42**, **43**, **47**, 50
Mk IIA (LR) 50
Mk IIB 43–4, **44**
Mk III 44, 50, 67
Mk IV (later XX) **83**, 50, 82–3
Mk IV/XX *see* Mk 21
Mk IX **9**, 58–9, 67–70, **70**, **72**, 73, 77–8, 82–3, **83**, 85–6, 88, 99, 101, 110–13, **110**
F Mk IX 69
FR Mk IX 99
HF Mk IX 69
LF Mk IX 69, 112
Mk IX-based Spitfire floatplane **78**, 79
Mk IXb **6**, **68**
Mk IXC
F Mk IXC **64–7**, **77**
HF Mk IXC **111**
LF Mk IXC **62**
Mk V 50–3, **54**, 55, 59, 60–2, **60**, **61**, 64, 67–9, 73, 76, 79, 83, 99, 101, 117–18
F Mk V 55, 56
LF Mk V 55–6
Mk VA 52, 55
Mk VB **6**, **48**, **50**, **51**, 53, 55, **58**, 59–61, **59–61**, 79, 117
Mk VC 52–5, 56–7, **56**, 58–61, 67, 68, 83, 110, **122**
Mk VI (high altitude, later the HF Mk VI) 57–8, **57**, 69, 72
Mk VII 64–7, 69, 70, 72, 101
F Mk VII **64**, 69
HF Mk VII 69
Mk VIII 59, 64–7, 68, 69, **71–3**, 72–6, **75–6**, 83, 85, 88, 101, 108–9, 110
F Mk VIII 73, **76**
HF Mk VIII 71, 73
LF Mk VIII 73
Mk XII **80**, 82, 83–4, **84**, **85**, 107, 121
Mk XIV 84–8, **86**, 106–7, 109
F Mk XIV **86**
FR Mk XIV 86, 89
Mk XIVE 86

FR Mk XIVE 86, **87**
Mk XVI 76–8, **78**, 112
LF Mk XVI **78**
Mk XVIII 88–9, **89**, 106, 111–12
FR Mk XVIII **88**, 89
Mk 21 **52**, 85, **104**, 106–7, **106**, **108**, 109, 123
Mk 22 **104**, 108, **108**, 123–4
F Mk 22 107–8, **107**
Mk 24 **104**, 108, **108**
F Mk 24 108
Mk392 **67**
PR Mk 19 **92**, 113
PR Mk IG **100**, **101**
PR Mk IV 82, **97**, 98–9
PR Mk IX 99
PR Mk Type 1A **93**, 94, **94**
PR Mk X (originally PR Mk VII) 99–101, 103
PR Mk XI **90**, 101–3, **101**, **102**
PR Mk XIX 101, 103, **103**, 112–13, **113**
PR Type G (later PR Mk VII) **98**, 99
pressurised 101
propellers 36–7
prototypes 9, 10–25, **10**, **21**, **25**
'Silver Spitfire' 113
'slipper' tanks 51–3
smoothness 23–4
Soviet 60–1, **60**
Spiteful (previously Victor Mk II) 108–9, 125
Spiteful F Mk XIV 109–10, **109**
Spiteful F Mk XV 110
T Mk IX 110
tail 24
tanks 48, 50, 51–4, 77, 96, 97, 98
Type 300 prototype (K5054) 6, **10**, **21**, 22, 24–5, **25**, 27–9, **28**, **29**
undercarriage 25
'wet wing' 98
Supermarine Type 224 (nicknamed Spitfire) 21–2, **21**
Swinton, Air Viscount 32

Taranto, Italy **73**
TE311 (Spitfire LF Mk XVI) **78**
Tempest Mk V 86, 88
Theseus (aircraft carrier) 125
'Thunderbolt' car 21
Tirpitz (battleship) 99, 120
'Triplex' rocket 78
Triumph (aircraft carrier) 123, **123**, 124–5, **124**
Tunisia **76**

Turkey 111
'turret fighter' concept 42

UB-403 (Seafire Mk XV) **121**
Unicorn (aircraft repair ship) 124–5
United States 8, 23, 77
United States Army Air Forces (USAAF) 59, **59**, 101
8th Air Force 101, **102**
12th Air Force 59
15th Air Force **77**
Groups
4th Pursuit Group 59
7th Photographic Group 101, **102**
31st Fighter Group 59, **72**, **73**
52nd Fighter Group 59, 73
Squadrons
14th Photographic Sqn 101, **102**
334th Pursuit Squadron 59
335th Pursuit Squadron 59
336th Pursuit Squadron 59
United States Navy 17, 117, 120
Cruiser Scouting Squadron Seven (VCS-7) 61
Unwin, George 'Grumpy' **35**

V-1 Missiles 84, 88
V-2 rocket sites 77
V-12 Curtiss D-12 engine 17
Vampire FB.5 **88**
Vichy regime 118
Vickers **10**, 23, **24**, 32, 44
Browning machine guns 28
Valiant nuclear bomber **24**
Wellington VI bomber 64–7
Vickers-Supermarine 110, 115
Vokes air filters **56**, 117
von Karman, Theodor 23
VP461 (Seafire F Mk 47) **124**

'washout' 24
Wasp (aircraft carrier) 53
Webster, Flt Lt Sidney 'Pebbler' Webster **18**
Weitner, Major Walter L 101
Western Desert 117
Westland **6**, 61, 119, 122
Lysander 92–3
Whittle, Flight Sergeant RH **75**
wing fence 24
Woolston 98
World Landplane Speed Record 34
Wright Field, USA 77

Yates, Warrant Officer WG **75**

Picture credits

AirSeaLand.Photos: 7–9, 14, 16, 18–22, 28–31, 36, 40/41, 45–50, 53, 59, 68, 70, 77, 83, 84, 91, 96–98, 109, 113, 114, 115, 118, 121, 125
Alamy: 13 & 17 (Chronicle)
Amber Books: 11, 26, 52, 57, 73, 74/75, 79, 81, 92, 93, 95, 101, 105, 108, 119, 123
Dreamstime: 63 (Andersonnancy)
Getty Images: 12 (Popperfoto), 24 (Keystone)

ARTWORKS
Amber Books: 33 all, 38, 42 top, 43, 44, 51 top, 54/55, 56 bottom, 61, 64, 65 all, 71 all, 86–89, 100 all, 102 all, 103 bottom, 106, 110, 111
Ronny Bar: 5, 25 both, 37, 42 bottom, 51 bottom, 56 top, 58, 66 all, 72 both, 76
Teasel Studios: 34, 35, 60, 67, 78 both, 85, 94, 103 top, 107, 112, 117, 120 both, 122, 124